Preface to the series

The study of environmental change is a major growth area of interdisciplinary science. Indeed, the intensity of current scientific activity in the field of environmental change may be viewed as the emergence of a new area of 'big science' alongside such recognized fields as nuclear physics, astronomy and biotechnology. The science of environmental change is fundamental science on a grand scale: rather different from nuclear physics but nevertheless no less important as a field of knowledge, and probably of more significance in terms of the continuing success of human societies in their occupation of the Earth's surface.

The need to establish the pattern and causes of recent climatic changes, to which human activities have contributed, is the main force behind the increasing scientific interest in environmental change. Only during the past few decades have the scale, intensity and permanence of human impacts on the environment been recognized and begun to be understood. A mere 5000 years ago, in the mid-Holocene, non-local human impacts were more or less negligible even on vegetation and soils. Today, however, pollutants have been detected in the Earth's most remote regions, and environmental processes, including those of the atmosphere and oceans, are being affected at a global scale.

Natural environmental change has, however, occurred throughout Earth's history. Large-scale natural events as abrupt as those associated with human environmental impacts are known to have occurred in the past. The future course of natural environmental change may in some cases exacerbate human-induced change; in other cases, such changes may neutralize the human effects. It is essential, therefore, to view current and future environmental changes, like global warming, in the context of the broader perspective of the past. This linking theme provides the distinctive focus of the series and is mentioned explicitly in many of the titles listed overleaf.

It is intended that each book in the series will be an authoritative, scholarly and accessible synthesis that will become known for advancing the conceptual framework of studies in environmental change. In particular we hope that each book will inform advanced undergraduates and be an inspiration to young research workers. To this end, all the invited authors are experts in their respective fields and are active at the research frontier. They are, moreover, broadly representative of the interdisciplinary and international nature of environmental change research today. Thus, the series as a whole aims to cover all the themes normally considered as key issues in environmental change even though individual books may take a particular viewpoint or approach.

John A. Matthews (Co-ordinating Editor)

Titles in the series

Already published:

Atmospheric Pollution and Environmental Change

Series Editors:

Co-ordinating Editor

John A. Matthews

Department of Geography, University of Wales, Swansea, UK

Editors

Raymond S. Bradley

Department of Geosciences, University of Massachusetts, Amherst, USA

Neil Roberts

Department of Geography, University of Plymouth, UK

Martin A. J. Williams

Mawson Graduate Centre for Environmental Studies, University of Adelaide, Australia

Atmospheric Pollution and Environmental Change

Sarah Metcalfe

School of Geography, University of Nottingham, UK

and

Dick Derwent

Department of Environmental Science and Technology, Imperial College London, UK

Hodder Arnold

A MEMBER OF THE HODDER HEADLINE GROUP

Distributed in the United States of America
by Oxford University Press Inc., New York

First published in Great Britain in 2005 by
Hodder Arnold, an imprint of Hodder Education, a member of the Hodder Headline Group,
338 Euston Road, London NW1 3BH
http://www.hoddereducation.com

Distributed in the United States of America by
Oxford University Press Inc.
198 Madison Avenue, New York, NY10016

Hodder Headline's policy is to use papers that are natural, renewable and
recyclable products and made fromwood grown in sustainable forests.
The logging and manufacturing processes are expected to conform to the
environmental regulations of the country of origin.

The advice and information in this book are believed to be true and
accurate at the date of going to press, but neither the author[s] nor the publisher
can accept any legal responsibility or liability for any errors or omissions.

British Library Cataloguing in Publication Data
A catalogue record for this book is available from the British Library

Library of Congress Cataloging-in-Publication Data
A catalog record for this book is available from the Library of Congress

ISBN-10: 0 340 71959 1
ISBN-13: 978 0 340 71959 6

1 2 3 4 5 6 7 8 9 10

Typeset in Palatino 10/11.5 by Pantek Arts Ltd, Maidstone, Kent
Printed and bound in Great Britain by J.W. Arrowsmiths Ltd., Bristol

What do you think about this book? Or any other Hodder Arnold title?
Please send your comments to www.hoddereducation.com

Contents

Figures

Acknowledgements

Sarah Metcalfe was inspired to write this book when she was unable to find a textbook covering the range of material that she wanted to teach her students. She would like to express her thanks to those students who went through the course as she was developing these ideas. The Institute of Geography at the University of Edinburgh provided some time to get the book started and the School of Geography at the University of Nottingham has enabled its completion. Sarah Metcalfe would also like to thank Dr Duncan Whyatt of Lancaster University, her long-time collaborator, for his patience while she has been distracted from other research tasks and for helping with a couple of the figures. Ben Vivian encouraged and cajoled and has lived to tell the tale.

Dick Derwent would like to thank Sarah Metcalfe for giving him the opportunity to work with her on this textbook as a testament to the almost twenty years that they have worked together on acid rain. He would also like to thank his wife, Brenda, and his son, Robin, for their inspiration and encouragement over many years.

Both authors would like to thank Chris Lewis (School of Geography, Nottingham) for preparing the vast majority of the figures. Finally, thanks to the series editors and Abigail Woodman at Hodder Arnold who have been very patient as the text edged towards completion.

1

Background

1.0 Introduction

Atmospheric Pollution and Environmental Change is a bold title for a book. No single volume can realistically attempt to address all the issues that might be included under this broad umbrella. Here the focus is on the composition of the atmosphere, how it changes through space and time, particularly as a result of human activities, and the effects these changes may have on humans, the natural (and built) environment and the global climate system. These changes are considered in the context of policy responses aimed at establishing standards (i.e. what concentration or deposition of a particular pollutant is deemed to be non-damaging and hence acceptable) and reducing emissions to bring the atmosphere and its surface exchange back towards something like a pre-disturbance state. There has been a tradition that people interested in air pollution issues were concerned largely with short timescales and local or regional (continental) spatial scales, while those working on environmental change looked to the larger picture. This division has been significantly eroded over the last 10 to 20 years, but is still preserved in many institutional structures, even in the definition of research programmes where air quality and climate change are separated. Here we attempt to bring together some of these different concerns over a range of spatial and temporal scales.

1.1 The atmosphere and its composition

The lower part of the earth's atmosphere (up to about 50 km from the earth's surface) contains about 99.9 per cent of the total atmospheric mass. It is the part of the atmosphere that dictates our climate and weather systems and supports life on earth as it is currently constituted. The main focus here is the troposphere, the bottom 10 to 16 km of the atmosphere (higher around the equator than at the poles), where the air is unstable due to surface heating and in almost constant motion. The composition of the troposphere (gases, particles and water vapour) determines both the pollution climate and the actual climate that we experience. Above the troposphere is the stratosphere, extending up to about 50 km above the earth's surface. Unlike the troposphere, this is a physically stable environment as temperature rises with height. The influence of the stratosphere on the troposphere is seen mainly through ozone which largely originates in the stratosphere, but which can be transported to lower levels under certain conditions. The influence of the troposphere on the stratosphere is probably better known, with long lived molecules (such as chlorofluorocarbons (CFCs)) ultimately making their way up into the stratosphere and causing seasonal depletion of the ozone layer.

This book will say almost nothing about the major chemical constituents of the atmosphere: nitrogen (N_2 – about 78 per cent of the atmosphere), oxygen (O_2 – about 21 per cent) and argon (Ar – about 0.9 per cent). N_2 is chemically inert and oxygen will feature here only as a source of the reactive ozone (O_3). Over the long term, the relative abundances of nitrogen and argon do not change significantly, but oxygen has shown a slight decline due to the burning of fossil fuels. The evolution of the earth's atmosphere from anoxic (dominated by carbon dioxide (CO_2)) to oxic (a high abundance of oxygen), which occurred mainly between 2 billion and 1 billion years ago, will not be part of this overview. Here the focus is on what are largely the trace constituents of the atmosphere, many of which are also highly reactive. Their concentrations can vary dramatically in space and time. In unpolluted air, the concentrations of these gases will be low, measured in units from ppm (parts per million) to ppt (parts per trillion). While the majority of these constituents have both natural and anthropogenic sources, there are a few (but significant) compounds that occur solely as a result of human activities. Many of the drivers of stratospheric ozone depletion are in the latter category. The major compounds described in this book are set out in Table 1.1.

The concentrations of these constituents of air may be expressed in a variety of units. For gases these will be as mole fractions expressed as volume mixing ratios (ppm, ppb etc) or as mass per unit volume (e.g. $\mu g\ m^{-3}$). Aerosols (solids or liquids) are expressed as mass per unit volume (e.g. $\mu g\ m^{-3}$ or $ng\ m^{-3}$). It is possible to convert between units such as ppm/ppb and $\mu g\ m^{-3}$, but assumptions have to be made about the molecular mass, pressure and temperature. Conversion factors for some of the common gases are given in Table 1.2.

1.2 Atmospheric lifetimes

The lifetime (or residence time) of a compound is a key factor to understand and will be determined by its chemical reactivity and how readily it can be removed from the atmosphere by dry or wet deposition (see also Chapter 4). The lifetimes of most of the common trace gases are measured in hours or days; the more readily they react, the shorter their lifetime. There are a few molecules which are essentially inert in the troposphere and which will slowly drift up into the stratosphere. Their lifetimes are measured in centuries or even millennia. Some of the best known of these long lasting compounds are those associated with stratospheric ozone depletion and they are discussed in more detail in Chapter 3. In some cases, such as the CFCs, it was their chemical stability that made them attractive for human use in the first place. Residence time can be taken as a rough indicator of spatial (and temporal) variability. Gases with a short residence time will tend to have high concentrations close to their sources, but these will then fall off rapidly with distance. By contrast, gases with long residence times will be transported through the atmosphere to become well mixed globally, producing an even distribution with altitude. Residence time is an important factor to take into account when assessing the likely impact of emissions control measures. Short-term benefits will be most apparent for gases with a short residence time.

Some trace gases, such as SF_6 and the perfluorocarbons, have exceedingly long atmospheric lifetimes of millennia and longer. These gases are a concern because they are 'immortal' molecules and make a permanent contribution to the greenhouse effect and climate change.

1.3 Transformation and loss

Chemical reactions in the atmosphere fall into two broad categories: thermal reactions and photochemical reactions. Thermal reactions are driven largely by the kinetic energy of the molecules involved (derived from their potential chemical energy), while photochemical reactions are driven by the energy from sunlight. These reactions will transform an initial set of species (reactants) into another set (products). Exceptions to this are the catalytic reactions where the catalyst is not consumed (lost) in the reaction process, but is maintained. Catalytic

TABLE 1.1 Major compounds discussed in this book and why they are of interest

Name	Symbol	Residence times	Area of concern
Nitric oxide	NO	1 day	U, A, C, E
Nitrogen dioxide	NO_2		
Nitrous oxide	N_2O	100 years	C
Nitrate aerosol	NO_3	Several days	A, E
Nitric acid	HNO_3	Several days	A
Ammonia	NH_3	Hours	A, E
Ammonium aerosol	NH_4	Several days	A, E
Sulphur dioxide	SO_2	Several days	U, A
Sulphate aerosol	SO_4	Several days	A, C
Carbon monoxide	CO	2 months	U, C
Carbon dioxide	CO_2	Decades/100s years	C
Methane	CH_4	12 years	C
Ozone	O_3	1 month	C, U
Lead particulate	Pb	Several days	U
Mercury	Hg	1 year	U
Cadmium particulate	Cd		U
Non-methane volatile organic compounds (e.g. benzene, toluene, ethane, polycyclic aromatic hydrocarbons)	NM-VOCs	Hours-days	U
Chlorofluorocarbons	CFCs	100 years	C
Hydrochlorofluorocarbons	HCFCs	10 years	C
Hydrofluorocarbons	HFCs	1 year	C
Perfluorocarbons	PFCs, CF_4, C_2F_6	10,000 years	C
Sulphur hexafluoride	SF_6	3000 years	C
Methyl chloride	CH_3Cl	1 year	C
Methyl bromide	CH_3Br	1 year	C
Hydroxyl radical	OH	1 second	G
Peroxy radical	RO_2	10 seconds	G
Hydroperoxy radical	HO_2	10 seconds	G
Hydrogen peroxide	H_2O_2	Several days	A
Particulate matter	PM	Several days	U, C

Key: U = urban, A = acidification, E = eutrophication, C = climate, G = general

TABLE 1.2 Conversion factors between ppb and $\mu g\ m^{-3}$ assuming a temperature of 20°C and 1013 mb pressure (Source: NEGTAP, 2001)

Gas	ppb $\rightarrow \mu g\ m^{-3}$ multiply by...	$\mu g\ m^{-3} \rightarrow$ ppb multiply by...
NO	1.25	0.80
NO_2	1.91	0.52
SO_2	2.66	0.38
O_3	2.00	0.50
NH_3	0.71	1.42
HNO_3	2.62	0.38

reactions are key in stratospheric ozone depletion (see Chapter 3). The rate at which a chemical reaction occurs is described by the reaction rate coefficient (k), which is often both temperature and pressure dependent and hence variable throughout the atmosphere. The chemical reaction rate is determined by both k and the product of the concentration in air of the reacting gases (i.e. the likelihood of molecules meeting). The products of photochemistry also need to be taken into account since sunlight-driven photochemical reactions drive much of the chemistry of the atmosphere. The major gas phase and photochemical reactions relating to particular pollutants are discussed in the following chapters. More details of the chemistry of the atmosphere (including aqueous phase chemistry) can be found in books such as Brimblecombe (1996) and Finlayson-Pitts and Pitts (1999).

Gases and aerosols can be lost from the atmosphere by the processes of dry deposition and wet deposition. The rates of these loss processes are often comparable in magnitude to losses through chemical reactions. From the point of view of the effects of pollutants on ecosystems and materials, deposition is a key process because it may lead to contamination and pollution of the surfaces. Rates of dry deposition are determined by air concentrations close to the surface and by deposition velocities (in m s). Deposition velocities are affected by aerodynamic diameter (for particles), roughness length of the surface, atmospheric conditions (especially wind speed), aqueous solubility (for gases) and biological activity for any vegetated surface. Whether surfaces are wet (or not) and whether plant stomata are open (biological activity or not) have also been found to be important in dictating rates of dry deposition. Dry deposition is rarely measured directly, but can be modelled (see Smith *et al.*, 2000). Wet deposition is the removal of gases or aerosols by precipitation. This can occur as rainout when the pollutants are incorporated into the cloud as particles which are then precipitated, or as washout when pollutants below the cloud are swept out of the atmosphere in the falling rain or snow. Wet deposition and related processes are discussed in more detail in Chapter 4.

1.4 Diffusion and dispersal

The distance that a pollutant will travel from its place of origin will be dictated by a range of factors including its rate of emission or production, the height of its emission (whether close to the ground or through chimney stacks), loss through chemical reaction and loss through dry or wet deposition. At the local scale, diffusion will be important as intense plumes of discharged gases and aerosols interact and mix with air molecules. Diffusion occurs along a gradient of concentration: concentrations are reduced with time and distance from source. Under normal atmospheric conditions, the diffusion and dispersion of a pollutant can be described using Gaussian plume (i.e. normal) distributions. These assumptions underpin the treatment of plume dispersion modelling, discussed further in Chapter 6. Turbulence and convection (driven by buoyancy, whether from natural heat from the sun, or from anthropogenic activities) also play key roles in dispersing pollutants. At larger scales, long-range transport is an important feature as pollutants are carried by the major wind systems of atmospheric circulation (e.g. the Westerlies in mid-latitudes). These advecting winds can carry pollutants hundreds or thousands of kilometres and often beyond the boundaries of individual countries and continents.

1.5 Pollutants and their sources

As described in Section 1.1, the vast majority of the constituents of the atmosphere have both natural and anthropogenic sources. What then is a pollutant? It is rather like a weed, something that is present in the wrong place, at the wrong time and in the wrong amounts. A formal definition of a pollutant is a trace constituent that, between the point of its discharge to the atmosphere and the point of its ultimate removal, causes damage to humans or the built and natural environment. Concentrations of gases or aerosols, or depositions of compounds, only become a source of concern if they can be associated with adverse effects. In the context of this book, these adverse effects will be associated with perturbations of the global atmosphere/climate system, damage due to acidic deposition or high levels of ground level ozone and adverse health effects in urban areas. Pollutants can be described as being either primary or secondary. Primary pollutants (such as SO_2 from coal combustion) are emitted directly into the atmosphere, while secondary pollutants (such as O_3) are the products of chemical or photochemical reactions. A few pollutants can be either primary or secondary. Nitrogen dioxide (NO_2) is a primary pollutant when it is emitted directly by motor vehicles and a secondary pollutant when it is produced by the reaction of NO and O_3.

A whole variety of processes will give rise to emissions of gases and particles (aerosols). These processes may be natural (e.g. respiration, volcanic eruptions) or anthropogenic (e.g. fossil fuel combustion in motor vehicles) or result from some combination of the two (e.g. biomass burning). They may also result from accidents, catastrophies or acts of sabotage and war. Understanding how much of each constituent of the atmosphere is emitted and where it is emitted from, is key in developing our understanding. The process of compiling emissions inventories is discussed in Chapter 2. Reliable emissions inventories are essential if computer models are to play a useful role in helping us to understand where and when pollution will occur and how it might be controlled. This is explored further in Chapter 6.

1.6 Pollution and change

The industrial revolution of the eighteenth and nineteenth centuries left the countries of Europe and North America with an all too visible legacy of smoky chimneys and urban grime. Poor air quality was apparent on the surfaces of buildings, peoples' washing and in poor respiratory health. Such concerns were generally local in scale. By the 1970s and 1980s, however, it was recognized that pollutants emitted in one place might well be having adverse effects many hundreds of km downwind. This pollution was not visible, but fish kills in streams and lakes and forest dieback were taken to be indicative of severe problems. Today, the focus has once again returned to the state of air quality in urban areas, although the pollutants involved vary between the different countries of the world. This local concern, however, is matched by anxiety about our impact on the atmosphere as a whole and the global climate system.

The global nature of air pollution problems became obvious during the 1950s with the advent of the atmospheric testing of nuclear weapons and the subsequent global scale radioactive fallout. The International Geophysical Year in 1957 brought the first measurements of atmospheric carbon dioxide and it soon became clear that the signal of fossil fuel burning from human activities was transported throughout the globe and ultimately to the South Pole. Other issues of global air quality are, however, much more recent and stem from concerns about the global scale build-up of greenhouse gases other than carbon dioxide. Such concerns initially surfaced during the 1990s as part of the assessment activities of the Intergovernmental Panel on Climate Change (IPCC).

This book looks at these issues at different scales. Chapter 2 describes some of the methods and networks available for assessing concentrations of pollutants now and in the past. Chapters 3, 4 and 5 focus on global, regional (i.e. continental) and urban pollution issues, but try to bring out how the same pollutant can play different roles in different places (scales). In each case, there is also an emphasis on the need to consider the time dimension. Understanding changes in atmospheric composition through time can be as important as understanding them through space. The use of computer models now underpins much atmospheric science and much air pollution policy, so the main approaches to modelling at different scales are described in Chapter 6, with case studies relating to Chapters 3, 4 and 5. Chapter 7 provides an introduction to the legislative frameworks that have evolved in relation to urban, regional and global air pollution. The final chapter (8) considers the relationship between science and policy and the need for both to come together to address air pollution issues successfully. A glossary of terms is provided to help the reader with the many acronyms and new terminology. Our ultimate conclusion is that no matter how good our scientific understanding, effective policy requires people and their politicians to come together.

Identifying atmospheric pollution

2.0 Introduction

Any study of the changing nature of the atmosphere (at any scale) requires data on the changing concentrations of atmospheric components across space and through time. The measurements may be direct (from ground-based networks, aircraft or satellites) or indirect (e.g. the records of long-term change preserved in ice cores). Ideally, networks of monitoring stations should provide high quality data across a range of parameters and through long time periods. In practice, networks are usually set up for specific purposes, with a limited budget, and monitoring sites may move or be closed down over time. Monitoring is expensive and tends to be concentrated in wealthier nations or areas (e.g. cities), while data for remote or poor areas are sparse. The potential for ground-based measurements over the oceans is clearly limited. It was only with the advent of satellite-based measurements that the capacity to make truly global measurements became a reality. As with all aspects of air pollution, the issue of air quality monitoring can be addressed at a range of scales. This chapter will review some of the networks attempting to provide data at global, regional (supra-national) and local (national) scales. The data from these networks, however, will never answer all the questions we have about changes in the composition of the atmosphere, particularly when we try to look forward into the future to consider the likely impacts of

emission reduction policies – or indeed a lack of them. Monitoring will have to be combined with activities such as the development of emissions inventories (discussed below) and modelling (see Chapter 6) (Bower, 1997).

2.1 Monitoring atmospheric composition

2.1.1 Monitoring at the global scale

Monitoring the state of the global atmosphere is clearly a huge challenge. The interests of individual nation states (the main source of funding) may not be most obviously served by setting up monitoring networks across the world. The practical difficulties of setting up and running networks of stations, with qualified staff and the appropriate analytical facilities, can be immense. Data collected by different organizations, but for a common goal, require a well-designed and implemented system of quality assurance and quality control, otherwise the resulting data will be of uncertain quality and of little practical use.

In 1989, the World Meteorological Organization (WMO) set up Global Atmosphere Watch (GAW). This brought together a number of longer running programmes including the Global Ozone Observing System (GO3OS) which started in the 1950s and the Background Air

Pollution Monitoring Network (BAPMoN) which started in the 1960s. GAW is basically a voluntary programme where the core funding comes from the members of the WMO and other participating countries. It links in to broader projects such as the Global Climate Observing System (GCOS) (http://www.wmo.ch/web/gcos/gcoshome.html).

The long-term objectives of GAW are:

- To provide authoritative scientific information and advice on the composition and behaviour of the global atmosphere and the factors that affect them.
- To establish and coordinate an operational system to determine global and regional levels and long-term trends of natural and man-made atmospheric constituents in order to forecast future states of, and stresses on, the environment and to enable governments to take prompt action to reduce pollution.
- To further the understanding of the chemistry and physics of the environment and climate-related atmospheric constituents and properties, and of the cycles of greenhouse gases in the earth system, and to apply this knowledge in the fields of meteorology and climatology, especially through the application of atmospheric models.
- To promote studies of the interaction of the atmosphere with the marine and terrestrial biosphere http://www.wmo.ch/web/arep/gaw_bground.html).

The GAW system comprises both global and regional networks. Global stations are located in areas remote from regional pollution sources (major cities, industrial complexes) and unlikely to be affected by natural pollution events such as volcanic eruptions or dust storms. These stations are designed to provide a view of the background state of the atmosphere in order to address global scale issues such as changes in total column ozone and other radiatively active gases implicated in climate change (see Chapter 3). Data collected include ozone (near surface, total column and vertical profile), concentrations of greenhouse gases (e.g. CO_2, CH_4, water vapour), UV, aerosol load, the chemical composition of rainfall, abundances of radioactive isotopes and general meteorological data. In 2004 there are 27 stations in the GAW global network (Fig. 2.1).

To ensure data quality, the WMO has prepared measurement manuals and has a system of quality assurance/quality control.

The GAW programme includes a series of World Data Centres (WDCs). The concept was originally introduced in 1957 for data from the International Geophysical Year, but has been developed to make a wide range of data available to the global research community. WDCs within GAW are:

- Greenhouse gases (Japan)
- Atmospheric trace gases (USA – Carbon Dioxide Information Analysis Center)
- Ozone and UV radiation (Canada)
- Aerosols (EU-Italy)
- Solar radiation (Russian Federation)
- Precipitation chemistry (USA).

The centres' websites provide access to a wide range of data and other information and can be accessed via the GAW home page.

Perhaps one of the most famous sites in the GAW global network is that at Mauna Loa in Hawaii (which is funded through the US NOAA Climate Monitoring and Diagnostics Laboratory). Measurements of CO_2 have been made at this site since 1958 and represent the longest continuous record available anywhere in the world. The data from 1958 to 2002 are illustrated in Fig. 2.2 (Keeling and Whorf, 2003) and show a clear upward trend from 315.8 ppm in 1959 to 372.95 ppm in 2002. The seasonal cycle in CO_2 concentrations is also evident. Data sets such as this have played a major role in identifying human impact on the global atmosphere as the site is in a location remote from major urban and industrial sources.

The regional element of the GAW network is focused on regional scale problems such as acid deposition, changes in tropospheric ozone and impacts of urban areas on the adjacent rural air environment (see Chapter 4). Regional stations are located to avoid the effects of local pollution sources and to be away from areas where rapid land use change is likely to occur. There are about 300 stations in the regional network, with data quality controlled by the use of calibration centres for the different pollutants and separate

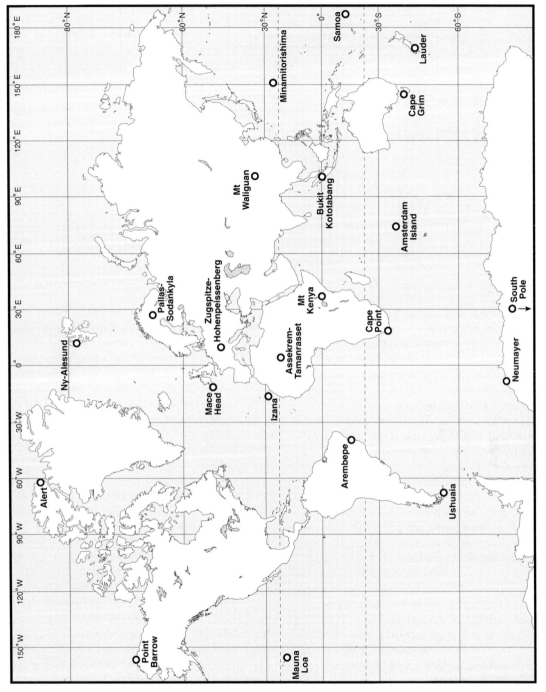

FIGURE 2.1 Stations in the Global Atmosphere Watch (GAW) network

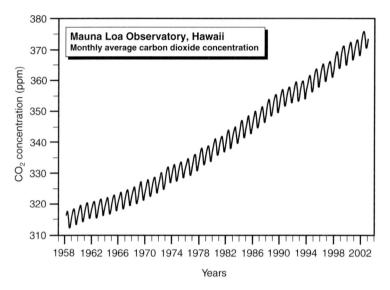

FIGURE 2.2 CO_2 measurements at Mauna Loa, Hawaii from 1958 (Data from Keeling, C. D. and Whorf, T. P. 2003)

quality assurance centres for Europe, Asia (and the Pacific) and the Americas. Data are fed into the GAW regional programme from other networks such as the European Monitoring and Evaluation Programme (EMEP), which is discussed in more detail below.

In the past GAW tended to develop in a reactive manner, and it became clear that it would be desirable to develop the programme in a more systematic and strategic way. As a result of this, a strategic plan was put forward to improve the quality and quantity of data available, to develop a 3-dimensional global network exploiting surface data, aircraft and satellite observations and to make the GAW data more widely available.

The most recent development in GAW has been the Global Urban Research Meteorology and Environmental Project (GURME) with a focus on urban air pollution (see Chapter 5). Some aspects of this project may be seen as taking over from a previous global urban air quality management programme (GEMS/AIR) run between 1975 and 1996 by the WHO (World Health Organization) and UNEP (United Nations Environment Programme).

International Global Atmospheric Chemistry (IGAC) is a core project within the International Geosphere–Biosphere Programme (IGBP).

IGAC focuses on biosphere-atmosphere interaction, oxidants and photochemistry and atmospheric aerosols (http://www.igac.noaa.gov/). There are also activities relating to education, developing global emissions inventories and modelling. IGAC was set up in the late 1980s, sponsored by the IGBP and the Commission on Atmospheric Chemistry and Global Pollution (CACGP). Its role is largely one of integration and synthesis, with individual projects being funded by national governments, but it has focused attention on the key role of the biosphere in determining the overall state of the atmosphere. The integration of biosphere and atmosphere is discussed further in Chapter 6.

Some monitoring programmes target specific problems. An example of this is the ALE/GAGE/AGAGE network which has monitored anthropogenic halocarbons associated with stratospheric ozone depletion since 1978, using high frequency gas chromatography. The instruments used for the measurements have changed over time, as have some of the monitoring stations, but a complete data set is available as the results of the earlier measurements (ALE – Atmospheric Lifetime Experiment and GAGE – Global Atmospheric Gases Experiment) have been recalibrated using the latest standards (AGAGE – Advanced GAGE). All the data are available through the Carbon Dioxide Information Analysis Center (http://www.cdiac.esd.ornl.gov/ndps/alegage.html).

There are currently five sites in the AGAGE network, located remote from major pollution sources (Tasmania, American Samoa, Barbados, Ireland, California). The closest sampling site to Europe is at Mace Head on the Atlantic coast of Ireland. Other gases monitored at these sites include methane (CH_4), nitrous oxide (N_2O) and carbon monoxide (CO). Data from these baseline stations have been used to track changes in the concentrations of ozone depleting chlorofluorocarbons (e.g. CFC–11, –12 and –113) and chlorocarbons (e.g. carbon tetrachloride (CCl_4)) and also the effectiveness of the Montreal Protocol and its amendments (see Chapter 7) in reducing these concentrations. They also offer the opportunity of using proxies for components of the atmosphere that cannot be measured directly at the large scale. A very significant example of this is the use of ALE/GAGE/AGAGE sites to estimate changes in the global concentration of OH (hydroxyl radical). This dominant oxidant in the lower atmosphere has apparently shown a general decline since the late 1980s (Prinn et al., 2001).

Halocarbon data from the AGAGE Mace Head site in Ireland (which is also a GAW site) have been analysed by Derwent et al. (1998). By sorting the data according to wind sector, it was possible to assign concentrations to either 'unpolluted' baseline air masses, or polluted air masses coming from the rest of Europe. The data indicate that by 1996 Northern Hemisphere baseline concentrations of all the major anthropogenic halocarbons, apart from CFC-12, had only begun to decline slowly (Fig. 2.3a). Concentrations in air masses originating in Europe, however, show very clear reductions over the period 1987 to 1996, reflecting cuts in emissions in response to the Montreal Protocol (Fig. 2.3b).

The measurement data do, however, raise some questions about the accuracy of estimates of European emissions of halocarbons. The issue of the complementarity between monitoring and emissions inventories is discussed further below.

No direct monitoring data are available which can provide continuous records extending back into the nineteenth century, and even less into the pre-industrial era. The gases and particles trapped in ice cores do offer the potential for such long-term records and have played a key role in identifying the broad relationship between concentrations of 'greenhouse' gases and the overall state of the global climate system (glacial or interglacial). The role of these ice core and firn records is discussed further in Chapters 3 and 4.

At the global scale, surface-based measurements are always going to be restricted. The ideal way to detect large scale change is to monitor over large areas and at a variety of heights in the atmosphere. Satellite-based systems offer this possibility and are becoming increasingly sophisticated. A good starting point is the NASA site http://www.terra.nasa.gov, with data from the Terra spacecraft earth observation system, which began collecting data in 2000. The earliest use of satellites was to

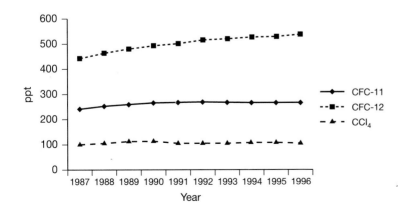

FIGURE 2.3a Apparent Northern Hemisphere baseline concentration of CFC-11, CFC-12 and CCl_4, measured at Mace Head, Ireland (Data from Derwent et al., 1998)

FIGURE 2.3b Concentrations above Northern Hemisphere background in air masses originating over Europe of CFC-11, CFC-12 and CCl_4, measured at Mace Head, Ireland (Data from Derwent *et al.*, 1998)

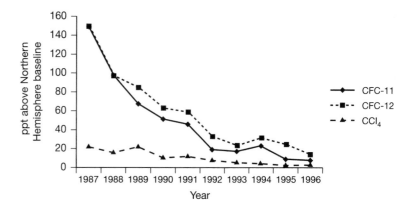

determine the composition of the stratosphere. Perhaps the best known of these systems is the Total Ozone Mapping Spectrometer (TOMS) which was first launched in 1978 on a NASA Nimbus-7 satellite. The original role of TOMS was to measure total column ozone and concentrations of SO_2 and aerosols resulting from volcanic eruptions. The TOMS ozone data (taken together with more limited surface-based measurements) have played a key role in tracking stratospheric ozone depletion. This issue is discussed further in Chapter 3. The latest platform for TOMS has been the Earth Probe satellite launched in 1996.

During the late 1990s, a big effort was made to refine the use of satellite data to study the composition of the troposphere in the context of total column values. This usually requires the separation of these two components by comparing values from 'clean' and 'dirty' regions at the same longitude (Tropospheric Excess Method – TEM). Chemical species with complex spectral signatures such as O_3 and NO_2 require further calculations. The European Space Agency launched the Global Ozone Monitoring Experiment (GOME) in 1995 specifically to study the lower troposphere. The main pollutants of interest are NO_2, SO_2, HCHO (formaldehyde), BrO (bromine monoxide) and O_3. Monitoring the composition of the troposphere from satellites has been reviewed by Borrell *et al.* (2000 and 2003). The Terra spacecraft carries an instrument called MOPITT (Measurements of Pollution in the Troposphere) which monitors carbon monoxide and methane.

Some of the earliest data on the troposphere came from TOMS for ozone (e.g. Fishman *et al.*, 1996). Satellite data have now been used to study issues such as biomass burning and the impact of forest fires, by combining data on aerosols and gas composition (especially NO_2, O_3 and HCHO). TOMS data (a 3-day composite) showing the smoke from fires over Indonesia in September 1997 are shown in Fig. 2.4.

The smoke resulted from a combination of both wild fires and the deliberate setting of fires during a period of extreme drought across the region associated with a strong El Niño. The link between ozone production and these fires has been described by Thompson *et al.* (2001). In the broader and more complex area of the global carbon cycle, satellites also have a role to play as they allow estimates of global primary productivity to be made. These estimates are based on monitoring changes in chlorophyll concentrations on the surface of the oceans and a vegetation index (Normalised Difference Vegetation Index – NDVI) over land (see http://www.seawifs.gsfc.nasa.gov/seawifs.html). Such methods can be used to monitor seasonal variations between hemispheres and the different responses of land and oceans and the effect of phenomena such as El Niño/Southern Oscillation. A recent study showed that while ocean productivity underwent major changes in the response to El Niño events, there was little impact on productivity on land at the global scale (Behrenfeld *et al.*, 2001). This difference in response was explained by the importance of areas of ocean upwelling for driving productivity in the tropical oceans.

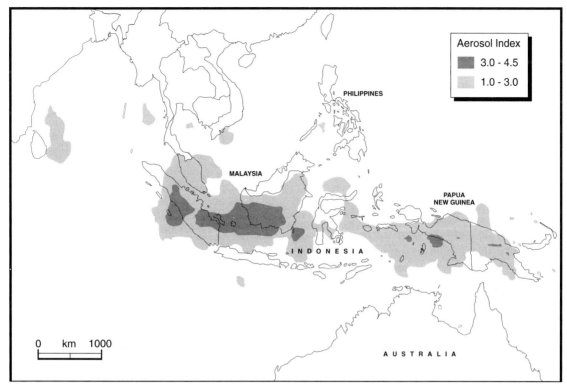

FIGURE 2.4 Smoke from fires over Indonesia in September 1997 based on TOMs data (Redrawn from NASA website)

2.1.2 Monitoring at the regional scale

As air pollution does not respect boundaries between countries (i.e. it is often a transboundary problem), so regional networks may be the most effective way to monitor patterns of air pollution. Sites are run as part of national monitoring programmes, but follow agreed protocols and reports cover the wider sampling area as a basis for policy making. In Europe and North America this approach has quite a long history, while in Asia it is a more recent development.

The European Monitoring and Evaluation Programme (EMEP) was created in 1976 with initial funding from the WMO and UNEP and has three main tasks: to compile emissions data, measure air and precipitation quality and develop atmospheric dispersion models. The programme is now funded by the signatories to the 1979 Convention on Long Range Transboundary Air Pollution (see Chapter 7).

EMEP monitoring began in 1978 with a network of 46 stations in 14 countries, but built on two earlier initiatives, the European Air Chemistry Network (1950s and 1960s) and an OECD programme in the early 1970s. The current EMEP network extends across 35 countries (Barrett *et al.*, 2000) and in 2001 had 140 sites monitoring a range of sulphur and nitrogen compounds and ozone, and 88 sites monitoring heavy metals and persistent organic pollutants (POPs). Sites monitoring the major atmospheric components are shown in Fig. 2.5.

Gases, particles and precipitation composition are all measured. Coordination of sampling (protocols for site location, sampling methods etc) and data analysis are carried out through the Chemical Coordinating Centre (CCC) housed at the Norsk institutt for luftforskning (NILU), Norway. Implementing quality assurance procedures is an important part of the CCC's role. Data from the EMEP network are

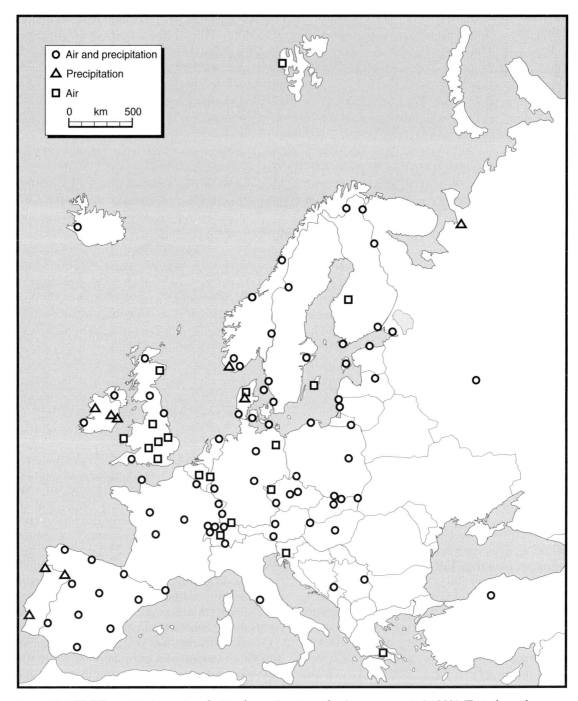

Figure 2.5 EMEP monitoring network sites for main atmospheric components in 2001 (Data from the Norwegian Meteorological Institute/EMEP/MSC-W)

available online for years 1995 onwards (http://www.nilu.no/projects/ccc/emepdata.html). Although the overall impression is one of gradual expansion of the EMEP monitoring network, the numbers and locations of sites in different countries, and what they measure, can be highly variable. For example, the measurement of nitrogen compounds, especially in air, only spread slowly through the 1980s. As a result, any assessment of long-term trends in air and precipitation quality has to be undertaken carefully. Such assessments are effectively the raison d'être for such networks, especially as they enable the impact of air pollution legislation to be assessed. Across Europe there have been two major reviews of long-term trends, one based on the European Air Chemistry Network (EACN) (Rodhe and Granat, 1984) and a more recent review based on the EMEP data (Barrett *et al.*, 2000). The latter showed clear downward trends in the concentrations of S in air and precipitation and in S deposition, from 1980 to the mid 1990s. The more limited N data (oxidised and reduced) show complex patterns. Changes in composition are consistent with changes in emissions. A wide range of data for Europe is available through the European Centre of Air and Climate Change (http://www.air-climate.eionet.eu.int) which has AirBase, an air quality information system.

A further European scale initiative has been The European Experiment on the Transport of Environmentally Relevant Trace Constituents over Europe (EUROTRAC). EUROTRAC started in 1986 and has focused on: ozone and other photo-oxidants in the troposphere, processes resulting in acidity in the atmosphere and the interactions between trace gases and the biosphere. As with most large scale projects, funding has come from member countries or, in this case, the European Union (EU). The first phase of EUROTRAC came to an end in 1995. A second phase ran from 1996 to 2002 and placed more emphasis on linking data and models to allow the development of more effective pollution control policies, particularly within the context of the United Nations Economic Commission for Europe (UNECE) and the EU (see Chapter 7). Monitoring projects within EUROTRAC have included studies of aerosol

and snow chemistry in alpine areas, cloud and fog processes and the use of satellites to monitor the composition of the troposphere. EUROTRAC is quite closely linked with EMEP and other European organizations, such as the European Environment Agency (EEA), and also has links with larger scale programmes such as IGAC.

In North America, the United States and Canada first recognized the need for a joint approach to pollution issues in 1909 (with respect to water quality), and the first air pollution agreement was associated with the Trail Smelter Arbitration in 1941. There is now a formal agreement between the two countries on air quality which is overseen by a committee of representatives from both countries. One aspect of such collaboration can be seen through the participation of Canada in the US National Atmospheric Deposition Program (NADP). NADP was set up in 1977 and now comprises three networks addressing weekly precipitation composition (NTN), daily precipitation composition (AIRMoN) and mercury deposition (MDN). (Canada finances sites in the MDN network at both federal and provincial level.) The entire NADP network now has some 200 sites (Lamb and Bowersox, 2000). Another collaborative project is the Integrated Atmospheric Deposition Network (IADN) which monitors air and precipitation composition in the Great Lakes basin. The network was set up in 1990 and aims to monitor inputs of toxic pollutants such as metals, PAHs, PCBs and pesticides. There is one 'master' monitoring station on each of the five Great Lakes and a series of satellite stations.

In Asia, the concept of regional networks is still in its infancy. The Acid Deposition Monitoring Network in Asia (EANET) started with a preparatory phase in 1998 and an official start in 2000. It began with an initial network of 38 sites (16 remote, 8 rural and 14 urban) across 10 countries, to monitor wet and dry deposition. The majority of sites are in China and Japan. Some of the EANET stations are also part of the GAW network. The administrative centre of EANET, called the Acid Deposition and Oxidant Research Center, is in Japan and performs a role similar to that of the CCC within EMEP. EANET also includes the monitoring of the effects of acid deposition on soils, vegetation and inland waters. Again in the EMEP mould, representatives of national governments will participate in decision making at the intergovernmental level. The EANET website can be found at http://www.adorc.gr.jp/index.html.

2.1.3 Monitoring at the national scale

The UK air quality monitoring network comprises more than 1500 sites measuring a wide range of pollutants in both urban and rural settings. There are a number of networks targeting different types of pollutants and providing records of different resolution (e.g. hourly, weekly) using different methods. Many of the sites in these networks are operated by local authorities or other organizations, but have central management and quality assurance procedures. The data from many of the UK's networks are available via the Internet (http://www.airquality.co.uk). Data from the networks for aerosols and ammonia concentrations are available from http://www.edinburgh.ceh.ac.uk/cara/networks.htm. The networks can generally be divided into two types: automatic and non-automatic (passive). The non-automatic sites use methods such as filter papers or diffusion tubes which have to be collected and analysed. Measurements at the automatic sites are made *in situ* and the data supplied directly to the management organizations via modem links. Reports on levels of air pollution in the UK are produced by the National Environmental Technology Centre (NETCEN) (e.g. Broughton *et al.*, 2000).

The non-automatic networks measure eight groups of pollutants:

1. **Smoke and sulphur dioxide (SO$_2$)** — the Smoke and SO$_2$ network is the oldest of all the national networks, although there are some records going back to before 1914. The national network (as the National Survey of Air Pollution) was set up in 1961 to track the implementation of the Clean Air Acts (see Chapter 7) and the sites are operated by local authorities. The original National Survey

had 500 sites, which increased to 1200 by 1966. In 1981, the National Survey became the UK Smoke and SO$_2$ monitoring network. In recent years the number of sites has declined to a present level of 165.

2. **Lead and other trace metals** — systematic lead measurements began in 1976, with a full set of data being available since 1980. This network was set up to monitor compliance with the EC directive on lead and the effects of the changing lead content in petrol. Results from a few of these sites are illustrated in Fig. 2.6 and they show a clear decline in the concentrations of lead in response to changes in the fuel.

3. **Heavy metals** — this network of 15 sites was established in 2003 to measure the concentrations of a range of heavy metals in rain and particulates. Unlike the previous heavy metal monitoring, the new network sites are in rural locations.

4. **Nitrogen dioxide** (using diffusion tubes) — the NO$_2$ diffusion tube network consists of more than 1100 sites in urban areas. Once again the sites are run by local authorities, but with quality assurance supplied by NETCEN.

5. **Acid deposition** — there are two elements to the Acid Deposition network: precipitation composition and rural gas concentrations (NO$_2$, SO$_2$). Again measurements of rainfall composition have a long history, with some data being collected in the nineteenth century (e.g. at Penicuik in central Scotland and at Rothamstead in southern England) (Brimblecombe and Pitman, 1980; Brimblecombe, 1987). When acid deposition emerged as a major environmental issue in the 1970s and 1980s, however, it soon became apparent that the data available were inadequate (RGAR, 1983). In 1984, the UK government recognized the need for a national network based on both daily (primary) and longer-term (secondary) sampling. The network underwent a major expansion through the 1980s such that by the time of the Third Report of the Review Group on Acid Rain (RGAR, 1990) there were 9 primary sites and 59 secondary sites operating over the period 1986–88. The network has subsequently been scaled back

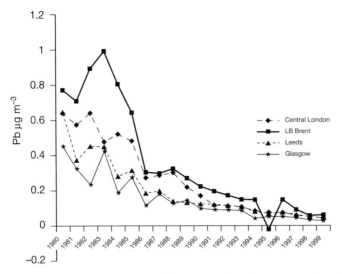

Figure 2.6 Annual mean lead (Pb) concentrations from UK multi-element sites since 1980 (Data from the UK Air Quality Archive, funded by DEFRA Scottish Executive, DoE N Ireland and Welsh Assembly Government)

and there are now 39 sites in the precipitation composition network and 5 primary sites, although of these, only Eskdalemuir continues to record daily data using a wet-only collector.

6. **Ammonia network** — this network was set up in 1996 to improve our understanding of a pollutant which shows a lot of variability through space and time. It has 95 sites using both active and passive sampling methods.

7. **Nitric acid monitoring network** — established in 1999 this 12-site network measures the concentrations of nitric acid and a range of other aerosols and acid gases. Prior to its establishment, measurements of aerosol concentrations were very sparse.

8. **Toxic Organic Micro Pollutants (TOMPs)** — the TOMPs network measures a number of organic, benzene-based compounds, many of which are known carcinogens. The major categories of pollutants measured are: polycyclic aromatic hydrocarbons (PAHs), polychlorinated biphenyl compounds (PCBs) and dioxins. There are currently 18 sites in operation, mainly in urban areas.

While the numbers of sites in many of the non-automatic networks have been in decline, there has been a substantial increase in the number of automatic sites. This reflects both increasing interest in urban air quality and the need to monitor pollutant levels to ascertain whether air quality standards are being met. There are more than 100 automatic sites (112 in 2000), the vast majority of which are in urban areas (Fig. 2.7).

Sites measure a range of gases, particles (PM_{10}) and hydrocarbons, but not all sites measure all pollutants. Some sites in the automatic network have been set up with funding from central government. Data from this network are available at very high resolution and can illustrate short-term variability (e.g. diurnal patterns in NO_2 associated with changes in traffic flows, see Fig. 2.8) as well as long-term trends.

These data play a major role in the identification of air quality problems associated with human health as many standards are set with respect to short exposure periods (see Chapter 7).

Data collected at the national scale are increasingly complemented by networks operated by local authorities. Although some urban areas have a long history of air quality monitoring, the Environment Act of 1995 requires all local authorities to review and assess air quality in their areas to ensure that targets set in the air quality strategy will be met (see Chapter 7). As a result of this legislation, monitoring has become both more widespread and more

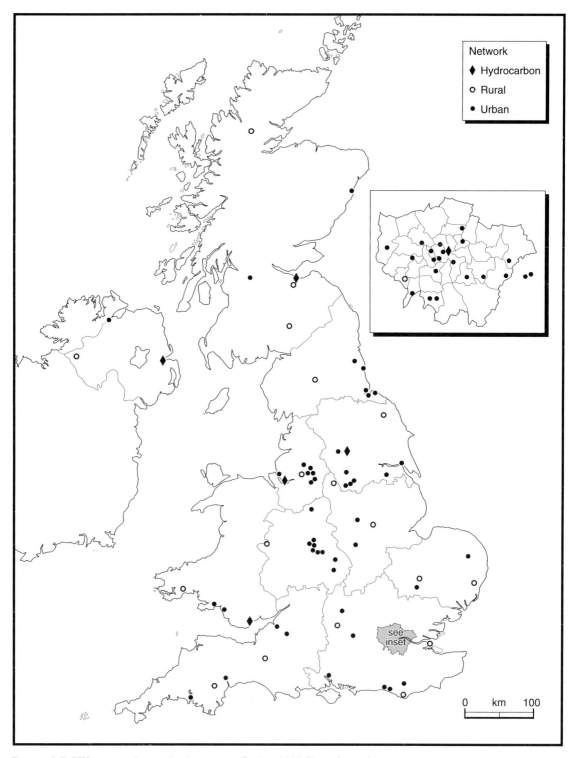

Figure 2.7 UK automatic monitoring network sites 2000 (Data from the UK Air Quality Archive)

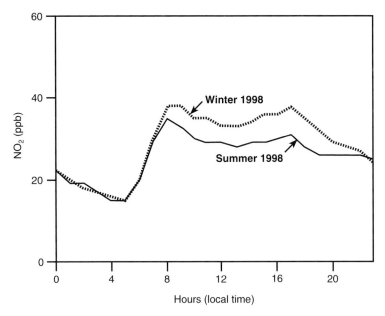

FIGURE 2.8 Hourly average NO$_2$ measured at Glasgow City Chambers (Scotland, UK) in 1998 (Data from the UK Air Quality Archive)

systematic, with clear guidelines being set down for the location of monitoring sites and recommendations for methods of sampling and analysis. This guidance is set out in a document produced by the Department for Environment, Food and Rural Affairs (DEFRA, formerly DETR) (DEFRA, 2000a) which is also available on the Internet (http://www.defra.gov.uk/environment/airquality/index.htm). It is a useful guide for anyone planning to set up an air quality monitoring network! The local authorities have to produce reports on air quality in their areas (a three-stage process), and these reports are often available through council websites. The stage 3 report for Edinburgh (City of Edinburgh Council, 2001) uses data from the national automatic urban network site, supplemented by data from the council's own automatic sampling programme (4 sites) and a network of 46 sites with passive diffusion tubes for NO$_2$. These higher spatial resolution data have enabled the City council to identify those parts of the city which they think will not meet air quality standards by 2005 (Air Quality Management Areas).

Information for London can be accessed through the London Air Quality Network site (http://www.londonair.org.uk/london/asp/home.asp). This site provides information on monitoring, standards, policy and air quality data by postcode and borough.

A different type of monitoring is provided by the Environmental Change Network (ECN). This was set up in 1992 in order to provide long-term monitoring of a range of physical, chemical and environmental variables. The initiative has been funded by a range of UK government departments and agencies, but there is no direct central government funding. The network comprises 12 terrestrial and 42 freshwater sites. Measurements at the terrestrial sites include NO$_2$ (measured using diffusion tubes) and precipitation composition. The network as a whole determines a whole range of water and soil chemistry and biological indicators such as vegetation, vertebrates, zooplankton and phyto-plankton (http://www.ecn.ac.uk).

Canada has two main air quality monitoring networks: the Canadian Air and Precipitation Monitoring Network (CAPMoN) and the National Air Pollution Surveillance (NAPS) network. There are other networks for mercury deposition (MDN – see above), air toxics and particulates. A brief introduction to these is available through the Environment Canada website (http://www.msc-smc.ec.gc.ca/ACSD/aqrb/index_e.html). NAPS was set up in 1969 as a joint initiative between federal and provincial governments with a view to providing air quality data for urban areas. The network provides information on ambient air quality in about 150 major population centres across the country and measures pollutants for which there are national ambient air quality objectives. In 1999, there were 252 monitoring stations. The monitoring sites are concentrated in the south of Canada (Fig. 2.9), particularly in southern British Columbia, Ontario and Quebec (Environment Canada, 2001).

As with the UK's automatic network, the focus here is on protecting human health, although the air quality objectives do cover broader environmental effects. There are clear quality assurance procedures and methods of analysis are also linked to standards in the USA. This last element is important as in provinces such as Ontario, the amount of pollution coming from US sources (especially related to ozone production) is a major concern and data have to be acceptable on both sides of the international frontier.

The role of CAPMoN is to monitor regional (i.e. non-urban) conditions and it currently has 19 sites in Canada and 1 in the USA. The site in the USA is to provide a point of comparison with the NADP network described above. The network measures precipitation chemistry, particles and trace gases and ground level ozone. CAPMoN itself began operating in 1983, but replaced two earlier networks which extend the data set back to 1978.

Canadian cities also operate their own air quality monitoring networks. Montreal has been monitoring air pollution since 1872. Since the broader Communauté Urbaine de Montréal was created in 1970, it has taken responsibility for air quality issues. Today the major focus is on ozone and fine particulates. The Communauté has a permanent network of 16 monitoring sites (Gagnon, 2001). Data from some of these sites feed into the NAPS (see above).

It is important to note, at this stage, that even air quality monitoring within an urban area will not give an accurate picture of the actual pollution to which an individual is exposed. Pollution levels in particular places (e.g. kitchens, bars) may be very different from the urban, or rural, average. The nature of indoor air pollution is discussed further in Chapter 5. The ultimate form of monitoring is personal exposure monitoring. Although this can be routine in certain work settings (e.g. for radiation exposure), for the general public it is generally limited to small sets of individuals, usually in the context of studies which are trying to relate exposure to pollutants to health outcomes. These limited data can, however, give an indication of the range of pollutant concentrations that individuals may actually experience compared with the levels indicated by routine monitoring programmes.

2.2 Emissions inventories

The compilation of emissions inventories, estimating emissions from different types of sources, is a complex process. The methods used to make these estimates vary between groups and over time, so that even the same organization may change its estimate of emissions for a given year. The number of actual measurements of emissions is extremely limited so there are considerable uncertainties in most estimates. Generally speaking, estimates from large, static point sources (e.g. power stations, industrial plant) are the most reliable, while those from natural area sources (e.g. soils) are the least reliable. As the dominant source sector is different for different components of the atmosphere, so estimates of emissions will vary in their uncertainty. In terms of anthropogenic emissions from the UK, for example, there is less uncertainty in the emissions of SO_2 (dominated by power stations), than in NH_3 (dominated by agricultural sources, primarily livestock). The error in SO_2 emissions has been estimated at 5–10 per cent (RGAR, 1997), while that for NH_3 has been estimated at ± 54 per cent (Sutton et al., 1995). For chemical species with longer atmospheric lifetimes, the broad validity of an

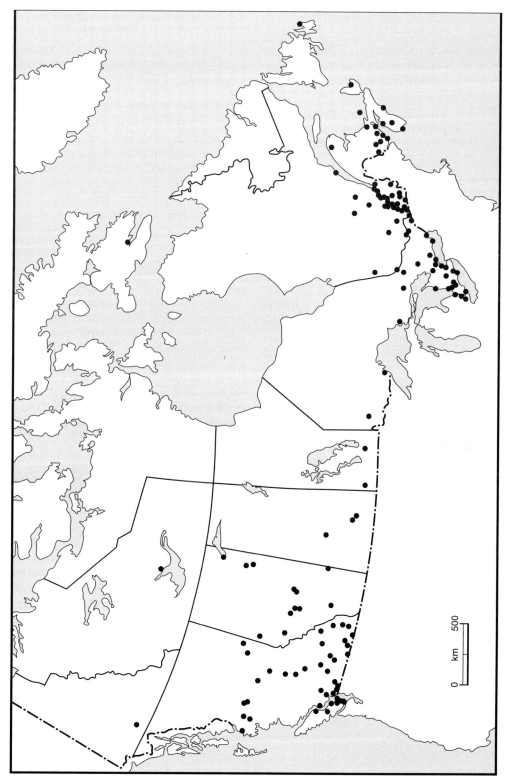

Figure 2.9 Sites in the Canadian National Air Pollution Surveillance Network (Environment, Canada, 2001)

emissions inventory can be established by comparing the inventory with data from monitoring networks. If areas shown on the inventory as having high emissions do not correspond to areas of high measured concentrations, then there is something amiss with the inventory!

Emissions inventories provide a useful point of comparison with monitoring data and, very importantly, provide information on the nature and distribution of sources which is invaluable in developing abatement policies. Successful policy depends on knowing what the major sources of pollution are and where they are. As with monitoring, there has been a trend towards trying to harmonize the methods by which emissions inventories are compiled, and there are now a number of international organizations which compile inventories using standardized methodologies. These include the UNECE, CORINAIR (Core Inventory of Air Emissions in Europe), EUROSTAT, the UNFCCC (UN Framework Convention on Climate Change) and NAPAP. While measured emissions data are increasingly available in developed countries for major point sources (e.g. power stations), emissions from other source categories have to be estimated. These estimates are based on a combination of an activity rate based on energy use (e.g. vehicle kilometres, or output of a particular industrial process) and an emission factor. Emission factors describe the relationship between the type and extent of the activity and the nature and quantity of the emissions, e.g. grams PM_{10} per km for different vehicle types. Emission factors will vary and will depend on fuel type and levels of technology; different emission factors will therefore apply in different countries and within countries over time. The United Kingdom has a set of emission factors (UK Emissions Factors Database) felt to be appropriate in the UK context and which can be accessed over the Internet (http://www.naei.org.uk/reports.php). In many countries no actual measurements will be available and inventories will rely entirely on information on fuel use and estimates of emission factors (not necessarily derived locally).

2.2.1 Developing an inventory

Emissions inventories can be derived using one of two major approaches: top-down or bottom-up. Top-down inventories will start with a statistic such as national fuel use (natural gas, fuel oil, coal etc) and then apportion that use based on the distribution of different source categories (often point, area and line sources), using information on the location of industrial processes, population densities and the road and rail network. In recent years this spatial allocation of emissions has generally been carried out within a Geographical Information System (GIS) framework. In a bottom-up approach, emissions are measured or estimated at the local level and then aggregated up. The latter approach is very time consuming and not generally applicable above the local scale.

Although emissions inventories for anthropogenic pollutants are uncertain, this uncertainty is small compared with trying to estimate emissions from natural sources. Developing emissions inventories for atmospheric components which have many natural sources and which are involved in complex cycles with different elements of the earth-atmosphere system (e.g. carbon) is particularly challenging, but has taken on major political importance in the context of issues such as global change and greenhouse gases. Uncertainties in emissions of greenhouse gases from the UK have been estimated to be in the range of ± 4 per cent for carbon dioxide to −55 per cent to +430 per cent for nitrous oxide (http://www.naei.org.uk). Estimates of uncertainties in other pollutants have been described by Passant (2003). It has now come to be recognized that we need the best possible estimates of emissions (and ultimately budgets) to address a whole range of issues from urban air quality to global climate change. In the sections below, the issues involved in compiling emissions inventories at local (urban), regional and global scales will be discussed using some specific examples. This discussion will be placed in a temporal framework, starting with estimates of present day emissions, then looking at reconstructing the past and finally developing emissions scenarios for the future.

2.2.2 Present day emissions

At the smallest scale, e.g. an individual car, chimney or chicken shed, it is theoretically possible to measure emissions. Within the UK emissions from large industrial point sources are now often measured. Even if they are not measured directly, it is still relatively straightforward to estimate emissions as detailed information is likely to be available about fuel type and use, combustion conditions etc and suitable emission factors can be employed. The Environment Agency in England and Wales (EA) and the Scottish Environmental Protection Agency (SEPA) in Scotland maintain an Inventory of Source and Releases for the sources for which they are responsible. The data from these point sources then feed in to the UK National Atmospheric Emissions Inventory (NAEI). Emissions from smaller point sources, mobile and area sources, are based on making a range of assumptions about fuel use and factors such as population, volumes of traffic flow, the composition of the vehicle fleet etc. Details of methods for estimating emissions are available in the DEFRA Local Authority guidance notes available on the Internet (DEFRA, 2000b).

Reports on the emissions of a range of pollutants implicated in acidic deposition, the generation of excess ground level ozone and climate change are also freely available through the NAEI (e.g. Goodwin et al., 2002). The most recent data are available through the NAEI section of the NETCEN website, including data from the national scale down to local authority level. Emissions are allocated to different source categories (e.g. road transport, agriculture, industrial combustion). These categories are often based on international agreements to ensure consistency between countries. There are, for example, agreed categories of emissions within the UNECE and the Convention on Long Range Transboundary Air Pollution. The harmonization of emissions estimation across the UNECE area is ensured by detailed guidance being provided by EMEP (see above) and CORINAIR working together under the umbrella of the European Environment Agency (EMEP/CORINAIR, 2004;

available via http://www.reports.eea.eu.int/emepcorinair4/en). There are slightly different categories for reporting emissions in the context of the UNFCCC. In this case, guidelines for reporting emissions come from the IPCC (IPCC, 1996). This IPCC methodology provides a common reporting base between countries. Thus the categories used by the US Environmental Protection Agency (EPA, 1999), China (e.g. Hongmin et al., 1996), Bangladesh (e.g. Ahmed et al., 1996) and the UK will all be the same.

While emissions data within individual countries may be available at high spatial resolution (e.g. the 1km data now available for the UK), at the regional level the resolution will be coarser. Across the EMEP area, most emissions data are now available at 50km resolution, although some countries still report emissions at the old EMEP resolution of 150km (see Chapter 4). The effect of the change in scale between the UK and EMEP for the map of NO_x emissions (as NO_2) is illustrated in Fig. 2.10.

On the UK map it is easy to distinguish the major conurbations and, indeed, the major arterial roads such as the M1 and M5/M6. At the EMEP scale, the concentration of emissions in the centre and south of England and in the central valley of Scotland is still apparent, but all the detail is lost.

Developing emissions inventories at the global scale is undoubtedly the greatest challenge, yet a task of fundamental importance if we are to improve our understanding of the earth–atmosphere system. The Global Emissions Inventory Activity (GEIA) is part of IGAC (described above) and can be accessed at http://www.geiacenter.org/index.html. The target year for their global inventories is 1990. It should be borne in mind that the uncertainties around some of these estimates are very large. Benkovitz et al. (1996) provide gridded global estimates of SO_2 and NO_x emissions for 1985, combining default global inventories with more reliable regional inventories where those were available (e.g. Europe – EMEP/CORINAIR, North America – NAPAP, Asia) (see Table 2.1). Emissions of sulphur come predominantly from human activities, with about

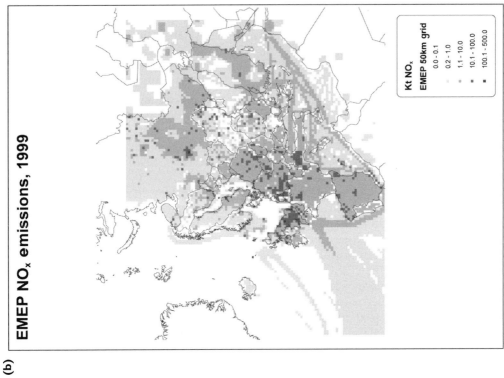

EMEP NO$_x$ emissions, 1999

(b)

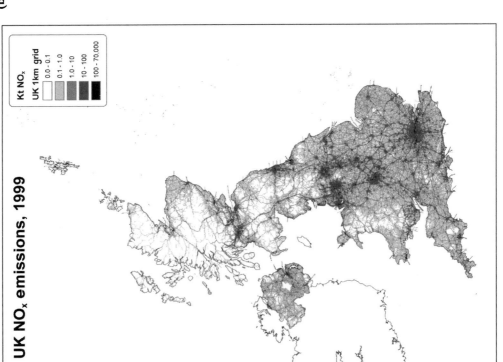

UK NO$_x$ emissions, 1999

(a)

FIGURE 2.10 1988 NO$_x$ emissions (in Kt) mapped for (a) the UK (1 km × 1 km) and (b) EMEP (50 km × 50 km) (Data from NAEI (UK) and the Norwegian Meteorological Institute/EMEP/MSC-W)

TABLE 2.1 Estimates of global S and oxidised N emissions for 1985 (in Tg S or N yr^{-1}) (Source: Benkowitz *et al.*, 1996)

	S	Range		N	Range
Anthropogenic	65.1		Anthropogenic	21	
Natural: Oceanic	15		Natural: Soil biogenic	5	
Terrestrial	0.4		Biomass burning	9	
Biomass burning	2		Lightning	10	
Volcanic	9				

80 per cent of the total originating in Northern Hemisphere mid-latitudes. Their best estimate of S emitted from anthropogenic sources is 65 Tg S yr^{-1}, while natural sources (on average) apparently contribute about 26.4 Tg S yr^{-1}. The largest natural source of sulphur is the ocean. The picture for the oxides of nitrogen (NO_x = $NO + NO_2$) is rather different, with a roughly equal split between anthropogenic and natural sources. The estimate for human activity is about 21 Tg N yr^{-1} and for natural sources between 15 and 20 Tg N yr^{-1}. Lee *et al.* (1997) estimate a total global emission of 44.3 Tg N. Of the natural sources, lightning and biomass burning contribute the most. Global oxidised N emissions are again dominated by the Northern Hemisphere mid-latitudes.

A global map of anthropogenic NO_x emissions is shown in Fig. 2.11 (from Benkovitz *et al.*, 1996) and can be compared with the EMEP and UK scale maps (Fig. 2.10). The effects of the change in scale ($1° \times 1°$ for the global map) are readily apparent.

Bouwman *et al.* (1997) developed a global gridded ammonia inventory for 1990. The total annual emission was estimated at 54 M tonnes of N of which the majority (21.6 M tonnes) came from farm animal waste. Total emissions of reduced N are greater than those of oxidised N and this trend looks set to continue.

Further information from around the world is available in Whelpdale and Kaiser (1996).

Improving the estimates of global emissions of radiatively active gases has been a major focus of the Intergovernmental Panel on Climate Change (IPCC). Their Third Assessment Report (TAR) (IPCC, 2001) adopts the estimates of total annual global emissions shown in Table 2.2. The equivalent figures from the Second Assessment Report (SAR) (IPCC, 1996) are also given. For CO_2 the figure is for emissions from fossil fuel and cement production only, while in the other cases both anthropogenic and natural sources are included.

Some of the uncertainties in emissions estimates for trace gases from natural sources or those related to land use (e.g. methane, nitrous oxide) have been discussed by Bouwman *et al.* (1999). As well as reviewing the methods available for estimating emissions, the authors point out that better spatial and temporal data (e.g. climate, soils, land use practices) are needed to yield inventories of the quality required by the kind of global chemical models which are increasingly part of global climate modelling. The topic of global scale modelling is discussed further in Chapter 6.

TABLE 2.2 Estimates of total annual global emissions of radiatively active gases (Source: Intergovernmental Panel on Climate Change, 1996 and 2001)

Emission	TAR (2001)	SAR (1996)
CO_2 (Pg C)	5.4 ± 0.3	5.5 ± 0.3
CH_4 (Tg CH_4)	598	597
N_2O (Tg N)	16.4	14.7
NO_x (Tg N)	51.9	N/A
		44.5 in IPCC (1995)

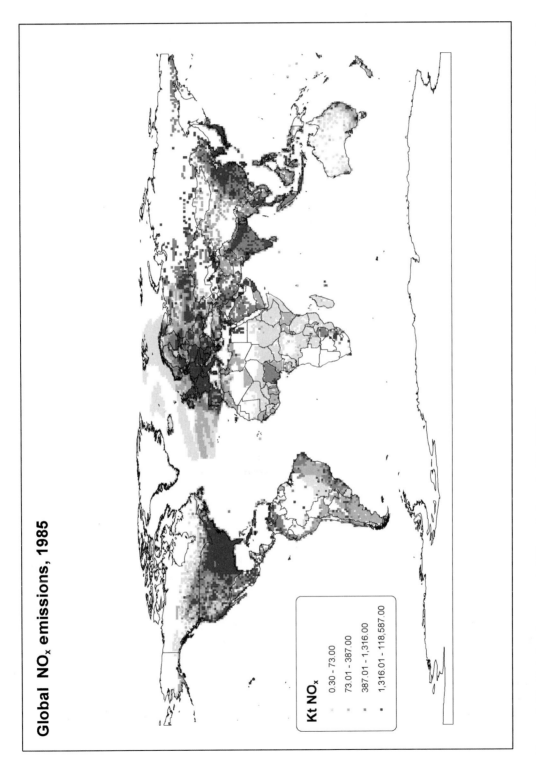

FIGURE 2.11 Global anthropogenic NO$_x$ emissions (in Kt) 1985 (Source: Benkovitz *et al.*, 1996 through the International Global Atmospheric Chemistry Programme (GEIA))

Earlier in the chapter it was pointed out that there are strong linkages between monitoring and emissions. It is possible to use measured concentrations to estimate emissions or fluxes; this is called inverse modelling. The measured pollution concentrations are compared with values modelled using current emissions data. If there is a discrepancy, then it may be that the emissions data need to be revised (although there may be other aspects of the model that are at fault). Mulholland and Seinfeld (1995) used this approach in a study of the South Coast Air Basin in southern California and were able to identify previously unrecognized areas of high carbon monoxide emissions. This approach has also been used to study global CO_2 fluxes, i.e. the changing balance between emission and uptake (see Chapter 6).

2.2.3 Estimating emissions in the past

As with monitoring, the adoption of a systematic approach to developing emissions inventories has been a relatively recent phenomenon. Across Europe and North America, most reliable inventory data begin in the 1980s. There are, however, occasional inventories available for earlier periods. In the UK, for example, there are isolated inventories for urban areas such as that produced for SO_2 from industrial sources in Sheffield by Garnett (1967) and there is a national SO_2 inventory for 1970 (Smith and Jeffrey, 1975).

It is also possible to apply modern methods to historical periods using the data available on industrial activity, fuel use, land use etc. These historical emissions estimates are inherently even less certain than those for the present day. Why then are these historical inventories valuable? In itself the changing geography of emissions is a useful guide to areas likely to suffer pollution problems. As the nature of sources changes, so do the types of environmental problems that might result. The inventories can be used in models to produce estimates of past concentrations and depositions of pollutants. These modelled estimates can be compared with the available monitoring data which can help to indicate whether the models are actually able to

reproduce the changing chemistry of the atmosphere (see Chapter 6). If the models perform reasonably well, then historical runs can be used to fill gaps in monitored data sets and indeed to provide estimates for periods before monitoring was in place. Historical estimates of emissions of SO_2, NO_x and VOCs have been compiled for the USA for a range of years since 1900 (EPA, 1992). Totals for CO and total particulates are available from 1940. Mylona (1996) provides information on SO_2 emissions across Europe from 1880; some of these data are illustrated in Chapter 4.

Although current methods of assessing damage from air pollutants (e.g. critical levels, critical loads) do not include a time element, it seems inevitable that the length of time over which stress or damage has been occurring will be important. As described in Chapter 4, for example, we have evidence of acidification of lake systems in north-west Europe from the mid-nineteenth century. There are no useful direct measurements of acid deposition at this time, but historical emissions inventories used in models indicate that S deposition would have been sufficient to cause acidification.

The quality of data available to compile global inventories for the historical period will clearly be highly variable both spatially and temporally. Considerable efforts have, however, been made to estimate both total emissions and their spatial distribution so that they can then be used in global scale models. Estimates of global CO_2 emissions from anthropogenic sources have been extended back to 1751 (Marland et al., 2003) and estimates for methane back to 1860 (Stern and Kaufmann, 1998). The trends in CO_2 are illustrated in Fig. 2.12 which shows a gradual increase through the nineteenth century, relatively stable emissions in the early twentieth century and then a very rapid rise.

Total methane emissions show a similar pattern. The change in amount has been matched by a change in spatial distribution (Andres et al., 1997). Since 1950, areas of high emission have spread over much of the Northern Hemisphere from the core areas of North America and Europe. The increase in south and south-east Asia is particularly notable, with increases of 63 per cent in China and 95 per cent

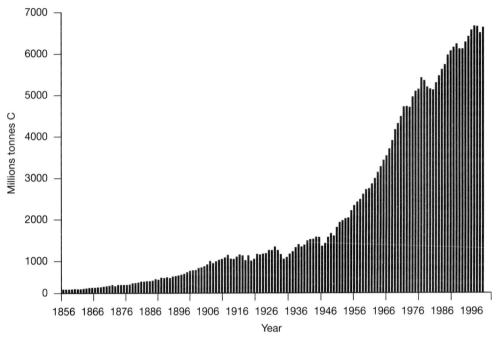

FIGURE 2.12 Estimated global CO_2 emissions 1850–2000 (Data from Marland *et al.*, 2003)

in India over the ten years from 1980. The picture is not one of increases everywhere, however, with some countries in north-west Europe showing a reduction in emissions during the 1980s. As the pattern of CO_2 emission reflects fossil fuel combustion, the pattern of global SO_2 and NO_x has followed a generally similar pattern (see Chapter 4 for further discussion). Estimates of pre-industrial emissions of these pollutants are often based on the simple measure of assuming zero fossil fuel combustion and a reduced input from biomass burning to reflect lower population densities.

2.2.4 Future emissions scenarios

While estimating emissions in the past can provide useful insights into atmospheric behaviour, developing future emissions scenarios is a primary part of policy making across a wide range of atmospheric pollution issues. Once again, the development of these scenarios is deeply bound up with modelling. Estimating future emissions will require a long-term view

of economic development (industrial, agricultural, social) as well as technological change. Increasingly we also have to take likely climate change into account and what effect this might have in certain areas, e.g. the use of central heating or air conditioning, and land use change. In the context of anthropogenic emissions, these future estimates will then also include the effects of deliberate emission reduction policies (see Chapter 7). Such policies will rarely (ever?) be based on obtaining the optimal environmental outcome, but will be a compromise between economic cost and environmental benefit. The economic aspect of emissions reduction strategies can be taken into account in Integrated Assessment Models (Schopp *et al.*, 1999). Any measure to reduce emissions is likely to be controversial at some level and policy will ultimately reflect a trade-off between what is technically feasible, economically possible and politically acceptable. Restricting traffic flow into city centres, or charging for access to improve urban air quality; limiting emissions from vehicles or factories to reduce acidifying pollutants; or agreeing targets for cuts in greenhouse gas emissions have all provoked debate and dissent.

Estimating changes in anthropogenic emissions, while posing challenges, is relatively easy when compared with estimating future emissions from natural sources. A whole range of variables including factors such as the occurrence of El Niño events (see section 2.1.1) or volcanic eruptions may play a part either directly or indirectly. We are increasingly aware that any attempt to estimate future emissions in order to model climate change also has to take into account the effect of climate change on emissions. These natural emissions in many cases have to be classed as uncontrollable, although there are a number of major greenhouse gases whose emissions could be manipulated through changes in land use practices.

While negotiations on emissions reductions within a country can be difficult enough, reaching agreement between nation states is even more complex. The longer the time frame and the broader the scope of the agreement, the more difficult the issue becomes. Even agreeing the possible emission scenarios becomes, in itself, a major political balancing act. Within the IPCC framework, there is a Special Report on Emission Scenarios (SRES) (IPCC, 2000). After much deliberation, the IPCC eventually used 40 possible scenarios which they grouped into four 'families' A1, A2, B1 and B2. Each 'family' makes different assumptions about global economic growth, population change and the introduction of new technologies. The effects of these assumptions on estimated global CO_2 emissions through to 2100 are illustrated in Fig. 2.13.

Estimates of total carbon dioxide emissions in 2100 range from a low of 4.23 PgC to a high of 29.09 PgC. The ranges for other pollutants are also large: from 20.2 TgS to 60.3 TgS for SO_2 and 236 $TgCH_4$ to 889 $TgCH_4$ for methane (IPCC, 2001).

In the short term, the impact of any emission scenario (i.e. its success in meeting the required objectives) can only be assessed by running the scenario in appropriate models. Some examples of the modelled impacts of emissions control strategies are given in Chapter 6.

2.3 Conclusions

Air quality monitoring provides the backbone of our understanding of the state of the atmosphere. While the overall level of monitoring

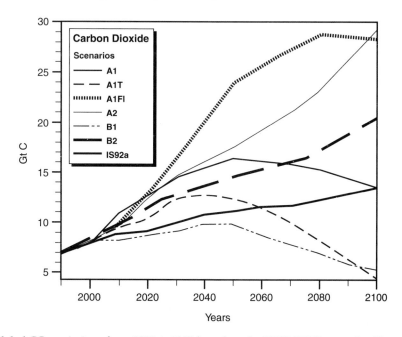

FIGURE 2.13 Global CO_2 emissions from 1990 to 2100 based on the IPCC SRES scenarios (Source: IPCC, 2001)

activity is undoubtedly increasing, it is still very limited when viewed from a broad spatial and temporal perspective. On the ground, networks change in response to changing political imperatives. Satellites obviously have great potential (Burrows *et al.*, 1990), but their data have to be validated against ground-based measurements. The composition of the air has only been directly monitored over a short period of time, although this can be expanded by the use of archives such as ice cores (see Chapters 3 and 4). Where (and when) we do not have monitoring data, we can use model output. These models need input in the form of emissions inventories, although some elements of these inventories can be highly uncertain. The Internet has revolutionized access to both monitoring data and emissions inventories from around the world. The following three chapters address a range of air pollution issues. The use of emissions inventories and scenarios comes to the fore again in Chapter 6 on modelling. Policy is reviewed in Chapter 7.

3

The global atmosphere

3.0 Introduction

One of the major advances of the latter part of the twentieth century was the recognition of the complex interrelationships between the chemical and physical composition of the atmosphere, the global climate and a set of feedbacks with the biosphere and hydrosphere. These relationships have been formalized in the term 'global biogeochemical cycles' and are represented in increasingly sophisticated ways in computer models which attempt to mimic the global climate system and its possible modes of behaviour (see also Chapter 6). Since the 1970s there has been an increasing focus on the ability of human activities to change the composition of the global atmosphere and, as a result, to induce changes in the global climate. We must not, however, lose sight of the fact that the global atmosphere has undergone massive natural changes in composition over both deep, geological timescales and over timescales of hundreds to thousands of years – typified by the last glacial cycle (ca. 150,000 years). This chapter focuses on two major issues: the enhanced greenhouse effect and the destruction of stratospheric ozone. Often viewed as separate areas of concern, they are in fact interrelated. Many of the components of the atmosphere discussed here in a global context are also important at the regional scale, where

concerns may relate to different types of effects (see Chapter 4).

3.1 The global greenhouse

Any discussion of the global atmosphere is inextricably bound up with the concept of the greenhouse effect – the trapping of long wave radiation from the earth's surface by gases and water vapour which keeps the surface of the earth and the lower part of the atmosphere some 33°C warmer than it would otherwise be. This natural process has been augmented by emissions of gases from human activities, giving rise to the enhanced greenhouse effect. The changes in the global climate system which seem likely to result from human induced changes in the atmosphere, have arguably become the major environmental issue of our time. The need for an agreed scientific basis to understanding climatic change, to assessing its likely impacts and to formulating response strategies resulted in the establishment of the Intergovernmental Panel on Climate Change (IPCC) in 1988. The IPCC was created as a joint initiative between the World Meteorological Organization (WMO) (http://www.wmo.ch or http://www.wmo.int) and the United Nations Environment Programme (UNEP) (http://www.unep.org).

3.2 Atmospheric composition and the climate system

The global climate system, and as a result the hydrosphere and biosphere, is driven by the energy coming from the sun. The amount and spatial distribution of this energy is affected (external to the earth-atmosphere system) by the geometry of the earth's orbit around the sun (astronomical forcing, typified by Milankovitch cycles) and by the changing output from the sun itself. The incoming solar radiation is in the form of short wave (largely visible) radiation which passes efficiently through the atmosphere. At the surface the incoming energy may be absorbed, or reflected back into the atmosphere as long wave (infrared) radiation (Fig. 3.1).

The amount of energy that is reflected back will depend on the reflectivity, or albedo, of the surface (water, ice, land – and if so the nature of the vegetation cover). Gases, water vapour, water droplets in clouds and aerosols (particles of different sizes) may absorb or scatter radiation on its passage through the atmosphere. This modification of the global energy balance is called radiative forcing. Long wave radiation is much more readily absorbed than short wave, hence a high proportion of the outgoing radiation from the earth's surface is effectively trapped in the atmosphere causing further warming. This is called the greenhouse effect – even though this is not how a greenhouse actually works. The efficiency with which radiation is absorbed at different wavelengths is illustrated in Fig. 3.2. This shows the high levels of

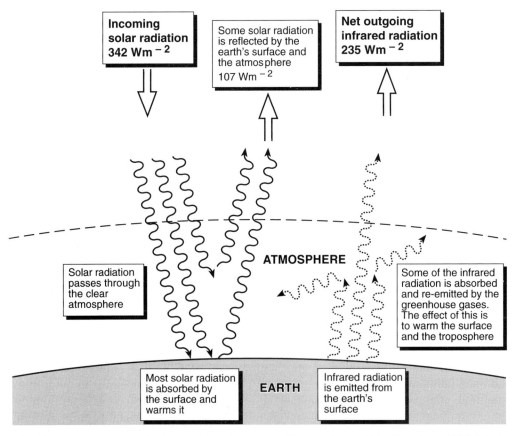

FIGURE 3.1 Long-term radiation balance of the atmosphere (Source: IPCC, 1994 (Fig. 7) and IPCC, 2001 (Fig. 1.2))

absorption associated with the major green-house gases (CO_2, water vapour). The role of CO_2 and water vapour in determining the heat absorbing capacity of the atmosphere was recognized by the nineteenth century British natural philosopher, John Tyndall.

There are, however, certain wavelengths where there is little natural absorption – these are called atmospheric windows. One window is in the short wave range (about 300 to 700 nm (0.3–0.7 μm)). Another window lies in the infrared region between about 8000 to 13,000 nm (8 and 13 μm) and provides the route by which most of the earth's outgoing radiation escapes back into space. This 'window' is not, however, completely open. There is an absorption peak associated with ozone and a range of additional compounds such as methane (CH_4), nitrous oxide (N_2O) and CFCs; HCFCs and other halocarbons also have peaks in this area. Growing concentrations of these gases, due to human activity for example, can effectively decrease the area of the open window. This effect can be exacerbated by increasing levels of water vapour in the atmosphere. The result of this positive radiative forcing is an enhanced greenhouse effect – global warming. The website of the Carbon Dioxide Information Analysis Center is a comprehensive source of information (http://www.cdiac.esd.ornl.gov/cdiac/home.html).

The effects of changes in CO_2 and water vapour on global temperatures were first quantified by the Swedish physical chemist Svante Arrhenius in two articles published in 1896 (Crawford, 1997). He calculated that CO_2 levels 1.5 times their mean would result in a 3.4°C increase in the global average temperature, while a decrease in concentration would lead to cooling. Although Arrhenius did consider the possible effects of increasing fossil fuel combustion on the future climate, his main interest was in explaining the changes in global temperature which gave rise to successive glacial and interglacial cycles. The significance of Arrhenius' work has been discussed in a special issue of the journal *Ambio* (Rodhe and Charleson, 1997).

Not all constituents of the atmosphere cause positive radiative forcing. Aerosols include a range of materials from dust and sea salt to secondary products relating to the oxidation of gases (see Chapter 4). The main focus of interest in climate studies has been sulphate aerosol which may come from large volcanic eruptions, the oxidation of sulphur produced by marine algae, or from the oxidation of SO_2 produced by fossil fuel combustion. Whether aerosols have a warming or cooling effect on the earth's surface depends on where they are in the atmosphere and whether they scatter or absorb radiation. Those low in the atmosphere which absorb radiation, tend to lead to warming, while those high in the atmosphere which scatter incoming radiation, lead to cooling. The general effect of an increased loading of aerosols (or particles)

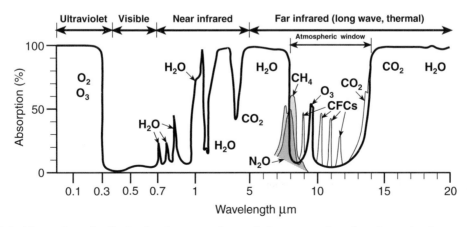

FIGURE 3.2 Absorption of radiation by the atmosphere relative to wavelength – absorption is expressed as the percentage of sunlight at the top of the atmosphere that is lost before reaching the ground (Based on Turco (1997) (Figs 11.10 and 11.12))

seems to be a negative radiative forcing since scattering generally outweighs absorption. Aerosols are also associated with the formation of clouds as they can act as cloud condensation nucleii. Clouds play a key role in influencing the radiation balance. Overall, clouds efficiently scatter solar radiation and result in a net cooling. In detail, however, the radiative forcing of clouds depends on their type (height, depth, water vapour or ice crystals). Low, dense clouds will lead to cooling at the surface, while high, thin clouds will lead to warming. Although the effect of clouds on radiative forcing is well understood in principle, clouds are extremely difficult to model (see Chapter 6), so they represent a large area of uncertainty in attempts to predict future changes in the climate due to human impacts on the atmosphere. The depletion of stratospheric ozone has been identified as another negative radiative forcing. This issue is discussed in more detail below.

The issues relating to radiative forcing and the mechanisms of the greenhouse effect (particularly the enhanced greenhouse effect) have been covered extensively in the literature. The links between the composition of the atmosphere and the rest of the climate system (incoming radiation, hydrosphere, cryosphere, biosphere) are key to understanding both past and future climatic change. Reports from the IPCC review the current scientific understanding (e.g. IPCC, 1996; 2001). The 1994 report of the IPCC focused specifically on radiative forcing (IPCC, 1995). The Third Assessment Report of the IPCC was published in 2001. Details of the IPCC's activities can be found on their website (http://www.ipcc.ch) or via the UK Meteorological Office (http://www.met-office.gov.uk/research/hadleycentre/ipcc/index.html).

The primary gases (excluding water vapour) involved in radiative forcing have been identified as: carbon dioxide (CO_2); methane (CH_4); nitrous oxide (N_2O); ozone (O_3) in both the troposphere (positive forcing) and stratosphere (negative forcing); and halocarbons and other halogenated compounds (CFCs, HFCs, PFCs). Aerosols in both the troposphere and stratosphere also play an important role (IPCC, 1996). The major sources and sinks of these compounds are described briefly below.

3.2.1 Sources of greenhouse gases and radiatively active aerosols

3.2.1.1 Carbon dioxide

CO_2 is the most abundant greenhouse gas in the atmosphere (excluding water vapour), with a current concentration of about 375 ppm. It occurs naturally and as a result of human activity. The major anthropogenic sources of CO_2 are fossil fuel combustion and deforestation. Natural fluxes are driven by the balance of photosynthesis and respiration, decomposition and cycling in the oceans. The C cycle is highly dynamic due to complex feedbacks between climate and the various sources and sinks. These feedbacks may be negative (such as increased CO_2 stimulating plant growth and hence encouraging CO_2 uptake) or positive (such as warming of the oceans reducing phytoplankton productivity and decreasing CO_2 uptake). CO_2 can be removed from the atmosphere by a range of processes into vegetation, soils and the oceans. The ultimate sink is the deep ocean, but CO_2 may be recycled from different stores back into the atmosphere over a range of timescales. Efforts to predict the future C balance therefore have to try to deal with processes operating over timescales ranging from years to millennia. Estimates of CO_2 atmospheric lifetime range from about 50 years to 200 years. IPCC (2001) carbon budget estimates for 1990–99 are:

$$\text{Emission (fossil fuel + cement)} = 6.3 \pm 0.4 \, \text{PgC/yr}$$
$$\text{Sinks — ocean/atmosphere} = -1.7 \pm 0.5 \, \text{PgC/yr}$$
$$\text{— land/atmosphere} = -1.4 \pm 0.7 \, \text{PgC/yr}$$

3.2.1.2 Methane

This gas also occurs both naturally and as a result of human activity; its current concentration is about 1.7 ppm. Wetlands (both tropical and high latitude) are the dominant natural sources of methane, probably contributing about 20 per cent of the global total. Anthropogenic sources include landfill, fossil fuels and agriculture, particularly livestock and rice cultivation. The magnitude of emissions from biospheric sources is still quite uncertain due to their high variability in space and time. The major sink for methane is through reaction with the hydroxyl radical (OH), with smaller losses to soils and to the stratosphere. The overall atmospheric lifetime of methane is estimated to be between 9 and 12 years, but varies considerably for different sinks. The feedbacks between CH_4 levels and the oxidising capacity of the atmosphere are discussed further in Section 3.4.2. The global CH_4 budget has been calculated by IPCC (2001) as:

Total emission = 598 $TgCH_4/yr$
(of which approximately 60% is anthropogenic)
Sinks — loss through
OH reaction = 506 $TgCH_4/yr$
— stratosphere = 40 $TgCH_4/yr$
— soils = 30 $TgCH_4/yr$

3.2.1.3 Nitrous oxide

This gas also has a range of sources, but emissions are poorly quantified. N_2O is produced by anaerobic processes in soils and by the oceans. Emissions from soils can be increased by farming practices, particularly the addition of N fertilizers. A number of industrial processes (such as nitric acid production) also emit N_2O. At the global scale it is believed that natural sources are more important than those resulting from human activity, although many of these 'natural' sources may be influenced by human activity (Bouwman *et al.*, 1995). The current concentration in the troposphere is about 312 ppb, compared with a pre-industrial level of 275 ppb (WMO, 1999). The main removal mechanism for N_2O is photolysis in the stratosphere, and as a result it has a long atmospheric lifetime of around 120 years. The budget estimates in IPCC (2001) are:

Emission — total = 14.7 Tg N yr
— anthropogenic = 6.9 Tg N yr
Sinks — atmospheric removal = 12.6 Tg N yr
— atmospheric increase = 3.8 Tg N yr

3.2.1.4 Ozone

Ozone is produced in the stratosphere (above about 20 km) as a result of photochemistry. UV light splits O_2 to produce free oxygen atoms, these oxygen atoms then react with O_2 and a mediator molecule (M) to produce O_3:

$$O_2 + uv\ light \rightarrow O + O$$
$$O + O_2 + M \rightarrow O_3 + M$$

Ozone is destroyed by photodissociation by UV and reaction with oxygen atoms. It is known that O_3 production shows little variability (perhaps 1–2 per cent during a solar cycle of about 11 years). Major changes in ozone concentration in the stratosphere are, therefore, a result of changes in O_3 losses (or sinks). Ozone can also be destroyed by catalytic reactions. Known catalysts for stratospheric ozone loss are chlorine, bromine, OH and nitric oxide (NO). As the catalyst itself is not destroyed in the reaction, even small concentrations of the catalysts can lead to significant levels of ozone destruction. Increased levels of ozone depleting species have been produced by human activity, especially over the last 50 years, and have resulted in severe losses of stratospheric ozone in high latitudes in the spring, the so-called 'ozone holes'. This is discussed in more detail in Section 3.4.

Some of the ozone produced in the stratosphere travels down into the troposphere. Within the troposphere ozone is produced by photochemical oxidation and destroyed at the earth's surface. The generation of excess ozone under polluted conditions is discussed in Chapter 4. Ozone concentrations in the lower atmosphere are highly variable through space and time. Production in the troposphere means that it acts as a traditional greenhouse gas, but losses in the stratosphere result in a negative radiative forcing (see Section 3.5). There is often confusion between the different roles of ozone in the stratosphere and the troposphere (Fig. 3.3).

In the troposphere the concern is over the presence of too much ozone, while in the stratosphere concern focuses on ozone loss. At the global scale the amount of ozone in the atmosphere is generally described in terms of total column ozone, expressed in Dobson Units (DU) after the British physicist George Dobson. Dobson developed a means of estimating the

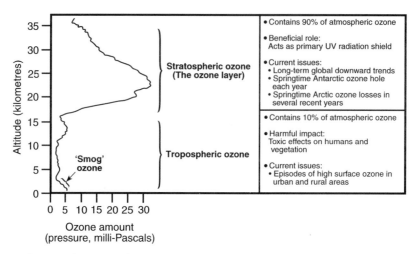

FIGURE 3.3 Distribution of ozone in the stratosphere and troposphere (Redrawn from p. xxvii of WMO (1998) with their permission)

concentration of ozone in a column of air of given dimensions based on the amount of UV light absorbed by the atmosphere. The global distribution of ozone shows a clearly zonal pattern, generally increasing from the tropics (ca. 250 DU) to the poles (> 300 DU). There is, however, much stronger seasonal variability at high latitudes than at low latitudes.

3.2.1.5 Halocarbons

These are carbon compounds containing chlorine, fluorine, bromine or iodine and they include chlorofluorocarbons (CFCs), hydro-chlorofluorocarbons (HCFCs), hydrofluoro-carbons (HFCs) and perhalogen species such as sulphur hexafluoride (SF_6). Halocarbons have been used as refrigerants, propellants and solvents since the early to mid-twentieth century. Many of them have very long atmospheric lifetimes (they have been called 'immortal molecules') and are very efficient greenhouse gases. The extent to which there are natural sources of these compounds is uncertain. Routine atmospheric measurements only began in the 1970s, by which time there was already a high loading from anthropogenic activities. Recent analysis of the gas composition of firn from cores in the Antarctic and Greenland (Butler *et al.*, 1999) suggest that there are no significant natural

sources except for methyl chloride (CH_3Cl) and methyl bromide (CH_3Br). The artificial halocarbons such as CFCs were designed to be chemically inert and hence, apparently, harmless. However, many are not biodegradable and have no significant sinks other than photodissociation by UV light in the stratosphere (as N_2O). This breakdown releases chlorine or bromine. Halocarbons containing chlorine and bromine (CFCs, HCFCs, halons) have been implicated in the depletion of stratospheric ozone, causing an indirect negative radiative forcing. Those halocarbons containing fluorine are associated with positive radiative forcing and global warming.

3.2.1.6 Aerosols

These can be derived from natural and anthropogenic processes and may be primary or secondary in origin. Natural sources of primary aerosols include sea spray, dust blown from dry continents and volcanoes. Anthropogenic sources include biomass burning and fossil fuel combustion. As described above, secondary aerosols such as sulphate are oxidation products. The atmospheric lifetime of aerosols in the troposphere is short (up to about four days), but those reaching the stratosphere may survive for a number of years. It should be noted that the production of SO_4 (and other aerosol species) is tied in with the ammonia cycle and ammonium aerosols are now the most abundant in the atmosphere.

3.2.2 Who emits what?

Since 1990, emissions of the main greenhouse gases in the USA have shown a largely upward trend, reaching 13 per cent above the 1990 baseline by 2001 (EPA, 2003a). The total emission in 2001 was 6936.2 $TgCO_2$ equivalents, which showed a small decrease (1.6 per cent) from 2000. The largest contribution was CO_2 from fossil fuel combustion, which accounted for 84 per cent of the total when each gas is weighted by its global warming potential (see Section 3.2.5). Most of the overall increase in the emissions of greenhouse gases from the USA can be attributed to CO_2 and has been caused by increasing petrol consumption and by a recent shift from natural gas to coal for power generation. Emissions from fossil fuel combustion increased by 17 per cent between 1990 and 2001. Emissions of CH_4 increased up to 1995, then declined. N_2O emissions, although declining since 1997, are still higher than in 1990. Emissions of HFCs, PFCs and SF_6 peaked around 1998 and are declining (but are still greater than in 1990). The increases in methane and nitrous oxide emissions were attributed to increasing emissions from motor vehicles, landfill and from more intensive land use. The persistence of relatively high emissions of HFCs, PFCs and SF_6 is due to their substitution for the severely ozone depleting CFCs (see below). These new compounds, however, have very high Global Warming Potential ((GWP) – see Section 3.2.5) and, in some cases, very long atmospheric lifetimes. The US EPA global warming website has up-to-date information and copies of reports (http://yosemite.epa.gov/oar/globalwarming.nsf/content/index.html).

Over the same time period (since 1990), UK emissions of CH_4 and N_2O have all declined. CO_2 showed a generally downward trend through the 1990s, but has increased in the early twenty-first century. Emissions expressed in IPCC common reporting format (CO_2 as C) have fallen from 164.8 M tonnes in 1990 to 156.1 M

tonnes in 2001 (Baggott *et al.*, 2003). Peaks in emissions occurred in 1973 and 1979, although the data were not expressed in the same way. During the early 1980s, emissions were low as a result of economic recession and in 1984 were severely affected by a national coal miners' strike (which also affected CH_4 emissions). The main source of CO_2 (35 per cent) is fossil fuel combustion for power generation, with 21 per cent coming from transport. Methane emissions also peaked during the 1970s. National emissions fell from 3.66 M tonnes in 1990 to 2.2 M tonnes in 2001. The major UK source of methane (42 per cent) is agriculture, mainly livestock. Waste disposal (landfill) contributes 24 per cent. Emissions of CFC substitutes have shown different trends, with SF_6 emissions increasing, while those of HFCs increased until 1998, but have since fallen to levels below those of 1990. Details of UK emissions are available through the National Emissions Inventory website (http://www.naei.org.uk/report_link.php). UK policy perspectives can be found on the DEFRA website (http://www.defra.gov.uk/environment/climatechange/index.htm).

3.2.3. Long-term archives of atmospheric change

Although the basic theoretical framework for the association between atmospheric composition and climate was in place by the end of the nineteenth century, empirical evidence of a long-term relationship between concentrations of greenhouse gases and global temperatures/ice volumes only became available in the 1980s. The retrieval of long ice cores, particularly from the polar regions, has provided archives of change which now extend back more than 700,000 years (EPICA community members, 2004). The practical difficulties of obtaining these cores, which can be more than 3000m long, should not be underestimated. The length and resolution of the ice core records depend upon the rate of accumulation at the coring site. High accumulation sites (warmer, wetter) give the highest temporal resolution, but are often limited in the total length of record. Low accumulation sites (colder, drier) have lower resolution but cover longer timespans. The longest records so far have come from Dome C and Vostok in the Antarctic (see above) and from central Greenland with the US Greenland Ice-Sheet Project (GISP2) and the European Greenland Ice-core Project (GRIP). Ice cores record the composition of the atmosphere in three basic ways: gases trapped in air bubbles; precipitation composition of the ice itself (chemical and isotopic); and dust loadings. The data from ice cores can be of value in studying atmospheric pollution at a range of scales and are also referred to in Chapter 4.

In 1980, scientists from the then Soviet Union began drilling an ice core at Vostok (3488 m a.s.l.) on the East Antarctic ice sheet. A series of cores have now been taken from this site yielding records of atmospheric change remote from any source of anthropogenic pollution. The original Vostok core provided a record of CO_2 concentration and global temperature change (based on δD) extending back into the penultimate glacial about 160,000 years ago (Barnola et al., 1987). The core showed a very close correlation between CO_2 concentration and isotopic temperature, with high levels of CO_2 during interglacials and low levels during glacials. This suggests a very strong radiative forcing link. CO_2 concentrations varied from < 200 ppm in full glacial conditions, to about 300 ppm at the peak of the last interglacial. The mean background concentration was found to be about 270 ppm, in keeping with estimates of pre-industrial levels. Glacial to interglacial transitions showed almost simultaneous changes in the two indicators, while the transition into the last glacial showed temperature decreasing ahead of CO_2. The authors suggested that this reflected the importance of forcings other than CO_2. The methane (CH_4) record from the same core showed a similar correspondence between high concentrations of the greenhouse gas and high temperatures. Methane concentrations of 0.35 ppm were recorded during the last glacial maximum and 0.65 ppm during the present interglacial (Holocene). Warmer periods (interstadials) during the last glacial were also marked by clear CH_4 peaks (Lorius et al., 1990). These authors attempted to assess what

proportion of the temperature change observed in the Vostok core could be explained by changes in greenhouse gas concentrations. Other factors taken into consideration were dust, non-sea salt sulphate, ice volume and local insolation. They found that between 40 and 65 per cent of the temperature change could be attributed to the effects of CO_2 and CH_4.

The overall relationship between greenhouse gas concentrations and Antarctic temperature was confirmed by results from the longer 400,000 year Vostok core of Petit et al. (1999) (see Fig. 3.4).

This record shows that current levels of CO_2, CH_4 and N_2O in the atmosphere (about 370 ppm, 1.7 ppm and 340 ppb respectively) exceed those at any time over the last 420,000 years. Interestingly, pre-disturbance concentrations in the Holocene were exceeded in earlier interglacials. The Holocene has been longer and climatically more stable than other interglacials. The data from the Vostok core are available at http://www.ngdc.noaa.gov/paleo/icecore/antarctica/vostok/vostok_data.html.

The question of leads and lags between changes in greenhouse gas concentrations and global temperatures is of clear interest given the current high and increasing levels of CO_2, CH_4, N_2O and other greenhouse gases. The use of ice cores to address this issue relies on having a sound chronological framework and confidence that differences between the age of the ice and the age of gas bubbles contained within it have been correctly identified. A study of the GISP2 core (Severinghaus and Brook, 1999) has shown that at the end of the last glacial, methane concentrations increased some 20–30 years after the initial warming occurred (Fig. 3.5).

The authors suggest that this delay reflects the time needed for increased CH_4 production to become established in the tropical wetlands. The importance of the tropics as a methane source has also been highlighted by results from an ice core from the Guliya ice cap in Tibet (Thompson et al., 1997). Data from the Greenland core can be obtained via http://www.ngdc.noaa.gov/paleo/icecore/greenland/greenland.html.

The main atmospheric components associated with negative radiative forcings are aerosols (particularly sulphate) and dust. Climatic effects

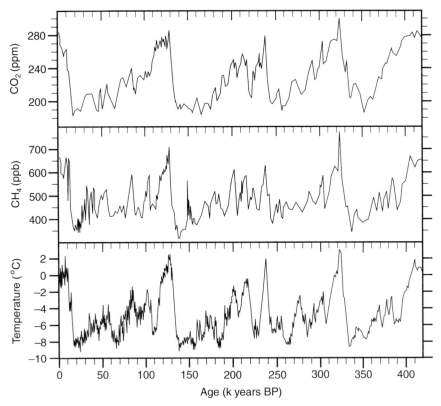

FIGURE 3.4 Changes in greenhouse gas concentrations in the Vostok core in relation to temperature over the last 420,000 years (Data from Petit *et al.* (2001) through the World Data Center for Paleoclimatology)

from increased dust loading have been particularly associated with volcanic eruptions, but also with the potential effects of nuclear weapons ('nuclear winter') and major fires from oil wells, such as those which resulted from the Gulf War in 1991. Dark smoke and aerosols both block incoming solar radiation and result in cooling at the surface. The possible link between the occurrence of volcanic eruptions and climatic cooling was first identified by Benjamin Franklin in 1784 when a very severe winter followed eruptions in Iceland during the previous summer. The eruption of Mount Tambora in Indonesia in 1815 was followed by the so-called 'year without a summer' of 1816. Studies of historical records of volcanic eruptions, climate data and records such as crop yields, resulted in Hubert Lamb's dust veil index (Lamb, 1970). Large particles and those finer aerosols which only get into the troposphere, have a short residence time in the

atmosphere and hence their effects tend to be localized and short term. Aerosols which reach the stratosphere, however, may persist for more than two years.

The volcanic eruptions which have the greatest effect on climate are those large eruptions in low latitudes which release abundant SO_2 high in the atmosphere where it is oxidized to SO_4. The aerosol produced stays in the atmosphere long enough to have a truly global impact (Robok and Mao, 1995). In recent times, the eruption which has had the greatest effect was that of Mount Pinatubo in the Philippines in 1991 (Fig. 3.6); the effects of this eruption on stratospheric aerosols persisted until 1997 (WMO, 1999).

The eruption resulted in a cooling of the global surface temperature of 0.3–0.5°C during 1992 and had a major effect on other aspects of atmospheric chemistry such as ozone concentrations in

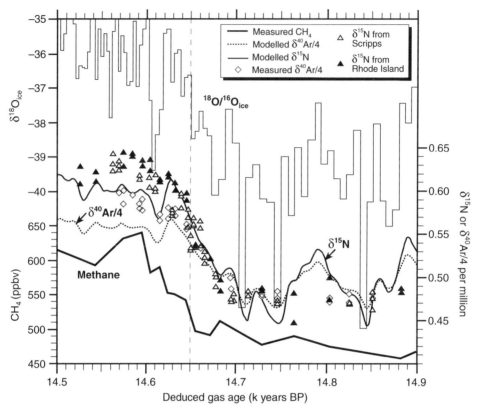

FIGURE 3.5 Temperature and methane change in the GISP$_2$ core around the Bølling transition (Redrawn from Severinghaus and Brook (1999), Fig. 3)

both middle and high latitudes (see below). It is believed that the eruption of Krakatau in 1883 had an even greater impact on the stratosphere. The possible impacts of both volcanic eruptions and solar variability on climate since 1700 have been investigated using a global climate model by Bertrand *et al.* (1999).

Ice cores provide long-term records of both these forcing components (positive and negative) of the atmosphere. Dust (microparticle) concentrations in ice cores are generally low during interglacial warm periods (such as the Holocene) and high during glacial periods. Compared with Holocene values, glacial dust flux may have been 10 to 30 times higher. High dust fluxes were associated with heavy metal inputs bound to the particle surfaces (Ferrari *et al.*, 2001). Increased dustiness has been associated with greater aridity on the continents,

lower sea levels and higher wind speeds. This is confirmed by records from loess deposits and deep sea cores (Fig. 3.7).

The increase in dustiness may have had a significant cooling effect, partly directly, but also through its association with fluxes of heavy metals. Increased iron deposition to the oceans may have increased productivity and hence increased CO_2 drawdown. The role of dust in the climate system has been the subject of increased interest, with explorations of both the past and the possible future (Harrison *et al.*, 2001). In 1998 a project called DIRTMAP (Dust Indicators and Records of Terrestrial Marine Palaeoenvironments) was established (http://www.ggy.bris.ac.uk/research/bridge/projects/DIRTMAP/dirtmap_main.htm) and a special issue of *Quaternary Science Reviews* has resulted (QSR 22, 2003).

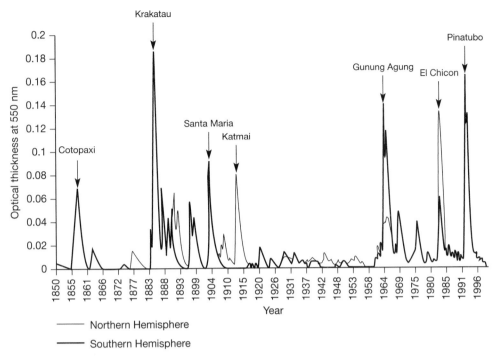

FIGURE 3.6 Observed stratospheric aerosol concentration and major volcanic eruptions (Optical data from http://www.giss.nasa.gov/data/simodel)

The fate of S emitted by anthropogenic activities is discussed in some detail in Chapter 4. However, naturally produced non-sea salt sulphate, mainly from oxidation of dms (dimethylsulphide) from marine algae, also shows high levels of variability over the long term. Ice core data from both the Dome C core in Greenland and the Vostok core from the Antarctic show that non-sea salt sulphate and MSA (methylsuphonic acid) concentrations increase sharply in glacial periods, compared with interglacials (a factor of 2 for non-sea salt SO_4 and a factor of 5 for MSA) (Legrand *et al.*, 1991). These increases have been attributed to increased algal productivity in the oceans during the cold, glacial stages, possibly due to enhanced upwelling. As sulphate aerosols are known to result in negative radiative forcing, there is obviously a complex relationship between climate and the natural S cycle. Short-term SO_4 peaks in ice cores are often associated with volcanic eruptions and the history of vol-

canic activity preserved in ice cores has been a focus of research (Hammer *et al.*, 1980; Zielinski *et al.*, 1994). Global volcanic activity seems to show some cyclicity (about 80 years according to Hammer *et al.*, 1980), so the climatic impacts might be expected to follow a similar pattern as the atmosphere will have less time to recover during periods of high activity. Although a great deal of interest has focused on the potential cooling effects of volcanic eruptions, the occurrence of high levels of volcanic activity during the last glacial (Zielinski, 2000) has also led to the suggestion that ice loading might trigger eruptions.

3.2.4 Recent change in the global atmosphere

For up-to-date information on concentrations of CO_2 and other greenhouse gases see http://www.cdiac.esd.ornl.gov/pns/current_ghg.html.

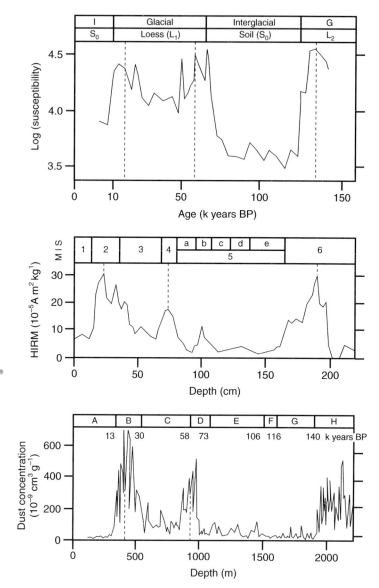

FIGURE 3.7 Dust records over the last glacial cycle from (a) the Vostok ice core (b) deep sea core S8–78 from the North Atlantic and (c) the Xifeng loess (Redrawn from Petit *et al.* (1990) Fig. 1, with permission. MIS = Marine Isotope Stage)

3.2.4.1 Carbon dioxide

Although quite a lot of measurements of atmospheric CO_2 concentrations were made in the early twentieth century, these data show a lot of scatter and are not believed to be reliable. Early evidence for increasing CO_2 concentrations was published by Callendar (1938). Changes in the ratio of the isotopes $^{14}C/^{12}C$ in wood cellulose, because of increasing atmospheric concentrations of old carbon from fossil fuels (from which the ^{14}C radioisotope had been depleted by radioactive decay), was published by Suess in 1955. Direct measurements of CO_2 began at the South Pole in 1957 and at the Mauna Loa Observatory, Hawaii in 1958. Since then several global CO_2 monitoring networks have been established, including one run by the US National Oceanic and Atmospheric Administration (NOAA) (http://www.cmdl.noaa.gov).

The data from the South Pole and Mauna Loa show that average atmospheric CO_2 concentrations have increased from about 315 ppm to more than 370 ppm (see Chapter 2). Over this period the difference between the Northern and Southern Hemispheres has also increased in response to the preponderance of fossil fuel emissions from the Northern Hemisphere. As described above, the ice core record shows that modern CO_2 levels are without precedent over the last 400,000 years, although diagrams which show a simple, inexorable rise since the end of the nineteenth century rather oversimplify the long-term complexity of the trends. The overall rise in atmospheric CO_2 since 1957 represents about half of the anthropogenic emission over that time period (Keeling *et al.*, 1995; IPCC, 2001). There was a slow down in the rate of increase in CO_2 concentrations in the early 1990s which might have been due to the effects of the Mount Pinatubo eruption. Any such effects, however, have been temporary and the late 1990s saw a very high rate of increase in atmospheric CO_2. Episodes of variability in CO_2 have also been attributed to the effects of the El Niño – Southern Oscillation. El Niño years are most often associated with low rates of CO_2 increase, but exceptions do occur and the 1998 El Niño saw the highest rate of CO_2 increase on record (IPCC, 2001).

3.2.4.2 Methane

Early CH_4 measurements were made in the nineteenth century. The pre-industrial background concentration was about 0.7 ppm (700 ppb). The more detailed measurements since the 1970s and particularly the 1980s, have shown that the rate of growth in atmospheric concentrations has been variable. Very rapid growth occurred during the 1970s (increasing at 20 ppb/yr), but then slowed to about 9 ppb/yr by 1991. The years 1992/3 were marked by very low rates of increase in CH_4 (or none at all in places); again this has been attributed to

perturbations caused by Mount Pinatubo (IPCC, 1996) and possibly reduced emissions from fossil fuel processing in the former Soviet Union. Growth has now resumed, although the slowing trend has continued (IPCC, 2001). Analysis of measurement data from remote sites for the period 1984–96 (NOAA-CMDL) has led to the suggestion that methane emissions may not actually be increasing (Dlugokencky *et al.*, 1998). The authors also suggest that there has been a shift in regional sources away from high northern latitudes towards the tropics. The rate of increase of N_2O has shown a similar pattern to that for CH_4, including the effects of the Pinatubo eruption. Since 1993, however, concentrations in the atmosphere have been increasing at levels as high as those of the 1980s (IPCC, 2001).

3.2.4.3 Low level O_3

Only limited measurements of tropospheric ozone are available away from the influence of urban areas. More discussion of early ozone measurements is given in Chapter 4. Those data which are available show considerable variation between different parts of the world. An analysis of a long-term data set from France shows that O_3 concentrations increased at a rate of 1.6 per cent per year from 1874 until 1980, although levels during the 1990s have been quite steady. At the Pic du Midi, ozone levels were 10 ppb in the nineteenth century and are now 48.5 ppb. WMO (1999) report surface ozone measurements from 17 sites around the world. The majority of sites show an increasing trend since the 1970s, although recent rates of growth are low. Europe has shown the biggest increase. Sites at the South Pole show a decrease. It appears that the available ozone data are consistent with what is known about the emissions of the major O_3 precursors (NO_X, NM-VOCs – see Chapter 4). Trace gas measurement campaigns have shown that ozone can be transported hundreds of km from its source region with plumes over the North Atlantic from the USA and over the Pacific from the adjacent Asian land mass (IPCC, 2001). In the Southern Hemisphere biomass burning during the dry season apparently makes a large contribution to seasonal ozone peaks. Modelling studies suggest that levels of tropospheric O_3 are likely to increase, which will have an impact on concentrations of the key oxidant, OH, and hence on the atmospheric lifetimes of other key gases such as CH_4 (IPCC, 2001).

3.2.4.4 Halocarbons

Tropospheric concentrations of halocarbons since the early 1990s clearly show the influence of a range of legislation introduced under the 1987 Montreal Protocol and its amendments (see Chapter 7). Growth of CFCs in the troposphere has slowed, stopped or declined (WMO, 1999; IPCC, 2001), following very rapid increases during the 1980s. Trends in the stratosphere seem to be similar, but with a lag reflecting the time taken for air to get from the troposphere and into the lower stratosphere (about one year). Levels of CFC production in the major industrialized nations are now below those of the early 1960s. Policy has driven a shift away from chlorine containing compounds such as CFCs and chlorocarbons (e.g. carbon tetrachloride (CCl_4)) and has resulted in a decline in organic chlorine concentrations in the troposphere. As might be expected, concentrations of CFC substitutes, the HCFCs and HFCs, have shown marked increases through the 1990s. HCFCs are, however, also controlled under the Montreal Protocol and will also be phased out between 2030 and 2040. Unlike CFCs, HCFCs and HFCs can be oxidized by OH in the troposphere and their oxidation products then washed out. The tropospheric lifetimes of most of these compounds is generally 10–20 years, much shorter than the compounds they have replaced. An exception is HFC-23 (a by-product of HCFC-22 manufacture), which has a lifetime of 260 years (IPCC, 2001).

Concentrations of halons (gases containing bromines) have continued to increase, although production in developed countries was due to end in 1993 under the terms of the Protocol. It is estimated that total organic bromine (CBr_y) is increasing in the troposphere at a rate of about 2.2 per cent per year. This is of considerable importance because, on a per atom basis, Br is extremely efficient in destroying stratospheric ozone (see below).

Concentrations of PFCs (perfluorocarbons) and SF_6 have both increased sharply through the 1980s and 1990s. The growth rate of SF_6 during the 1980s and 1990s was about 7 per cent per year. Both of these are very efficient greenhouse gases, but do not play a part in the destruction of stratospheric ozone. A new hybrid of PFCs and SF_6, called SF_5CF_3 (trifluoromethyl sulphur pentafluoride), has recently been discovered (Sturges *et al.*, 2000) which seems to have the largest radiative forcing potential of any gas found so far. This only seems to have been present in the atmosphere since the late 1960s and has an estimated lifetime of about 3200 years!

3.2.5 Making comparisons

In order to make it easier to compare the effects of the different greenhouse gases on the global energy balance, the concept of the Global Warming Potential (GWP) has been adopted. The index defines the radiative forcing of a unit

TABLE 3.1 GWPs for selected greenhouse gases (Source: IPCC, 2001)

	Lifetime (yrs)	GWP 20 years	GWP 100 years	GWP 500 years
CO_2	50–200, but variable	1	1	1
CH_4	12	62	23	7
N_2O	114	275	296	156
HFC-23	260	9400	12000	10000
SF_6	3200	15100	22200	32400

emission of a given gas relative to that for CO_2 (used as the reference) assuming that the background atmosphere remains constant (IPCC, 1996). GWPs have been estimated for many of the greenhouse gases, but not for aerosols due to their patchy distribution through the global atmosphere. The GWP of a gas can be calculated over a range of timescales. GWP estimates have to include both direct and indirect effects, for example ozone depleting substances tend to cause a positive forcing as a direct effect, but a negative forcing as an indirect effect through depletion of stratospheric ozone. Some typical GWPs, assuming constant CO_2, are given in Table 3.1. It should be borne in mind that, at best, these GWP estimates are likely to be accurate to within ± 35 per cent, such is the complexity of the system.

3.3 Stratospheric ozone

Detailed assessments and descriptions of the issues relating to stratospheric ozone are given in WMO reports (1999, 2003); the key chemical processes are also reviewed in Solomon (1999). There are a number of websites offering both good background and up-to-date information (http://www.atm.ch.cam.ac.uk/tour/index.html; http://www.ace.mmu.ac.uk/resources.html; http://www.ozonelayer.noaa.gov/ the World Meteorological Organization http://www.wmo.ch/indexflash.html (see ozone bulletins and maps); the US EPA http://www.epa.gov/docs/ozone and the Network for the Detection of Stratospheric Change http://www.ndsc.ncep.noaa.gov. There is also a textbook on stratospheric ozone available online (http://www.ccpo.odu.edu/SEES/ozone/oz_class.htm). With this wealth of information available, only the major issues will be discussed here.

About 90 per cent of all the ozone in the atmosphere occurs in the stratosphere, mainly between 8 and 18 km above the surface. This is known as the ozone layer. It was recognized in the late nineteenth century that stratospheric ozone effectively blocks a large proportion of incoming UV-B radiation in the range 280–320 nm. Excessive exposure to UV-B has been associated with skin cancer, cataracts and suppression of the immune system. Concerns about the destruction of stratospheric ozone by chlorofluorocarbons (CFCs) were first voiced in the 1970s (Molina and Rowland, 1974) and resulted in the USA banning their use in aerosols as early as 1979. There were also worries that a new generation of supersonic aircraft, emitting NO_X directly into the stratosphere, would also deplete ozone. The British Antarctic Survey (BAS) had been measuring total column ozone since 1957 and first recorded a decline in the early 1980s. In 1985, BAS first described an ozone 'hole' occurring in the Antarctic spring, with a loss of half the total column ozone, mainly between 13 and 22 km above the surface (Farman *et al.*, 1985). By 1988, scientists became curious as to whether similar depletion occurred in the Arctic spring. An expedition found that similar chemistry was occurring in the high northern latitudes and the study was extended in 1991–92. It is clear that depletion occurs in the high latitudes of both hemispheres, although it is more intense around the Antarctic.

Ozone loss in the stratosphere is driven by catalytic cycles involving chlorine (Cl) and chlorine monoxide (ClO) radicals and bromine (Br) and bromine monoxide (BrO) radicals. The catalytic cycle is:

$$O_3 + Br \ (\text{or } Cl) \ \rightarrow BrO \ (\text{or } ClO) + O_2$$
$$BrO \ (\text{or } ClO) + O \rightarrow Br \ (\text{or } Cl) + O_2$$

The overall result is that the odd oxygens O_3 and O are replaced by $2O_2$.

Under normal atmospheric conditions, only about 1 per cent of chlorine in the stratosphere is in this reactive form, while the percentage of reactive bromine is higher. At high latitudes, however, the reactive species of chlorine and bromine are produced by fast, heterogeneous reactions on the surface of ice crystals in polar stratospheric clouds (PSCs). PSCs can only form under intensely cold conditions (< –80°C) and as a result are more common in the Antarctic than in the Arctic. The reactions in the PSCs also inhibit the production of non-reactive species of Cl and Br. The enhanced abundance of reactive Cl and Br species resulting from these reactions with PSCs leads to greater O_3 loss. In the warmer polar regions and in mid-latitudes, sulphate aerosols may perform a similar role to PSCs. In the presence of sunlight, the reactive chlorine and bromine can destroy ozone; bromine can also destroy ozone without sunlight. The extreme cold over the Antarctic, the abundance of PSCs and its isolation from the rest of the global atmosphere, as a result of the circumpolar vortex, make conditions in the spring ideal for O_3 depletion. The Arctic is less cold and the atmosphere less isolated, but depletion still occurs. It seems that bromine may play a more important role in the Northern Hemisphere than it does in the Southern Hemisphere. Nearly all the chlorine and about 50 per cent of the bromine in the stratosphere come from human activities. There is also a significant natural source of methyl bromide (CH_3Br), contributing about 70 per cent of the total emission.

As with the greenhouse gases, it is useful to be able to make direct comparisons between the efficiency of ozone depleting substances. This has resulted in the definition of Ozone Depletion Potentials (ODPs). Here the change in O_3 per unit emission of the different compounds is compared with the change from a unit emission of CFC-11. The latest ODPs have been published in WMO (2003). Some are listed in Table 3.2 and show both the most recent estimates and those assumed in current legislation.

3.3.1 Trends in stratospheric ozone

Measurements of total ozone have been made using satellites since 1978. The Total Ozone Mapping Spectrometer (TOMS) has been built and operated by the US National Aeronautics and Space Administration (NASA), with a series of instruments providing records from 1978–1994 and since 1996 (see Chapter 2). The first major review of TOMS data was published by Stolarski *et al.* (1991). TOMS output is widely available through the Internet (http:// www.toms.gsfc.nasa.gov or http://www.badc.nerc.ac.uk/data/toms/index.html). The satellite data are complemented by ground-based and aircraft measurements. Results are expressed in Dobson Units (DU) (see Section 3.2.1.4).

Total ozone concentrations show a strong response to the seasonal cycle, the Quasi Biennial Oscillation (a reversal of wind direction in the tropical stratosphere, accelerating or decelerating flow), the 11-year solar cycle, ENSO, the NAO (North Atlantic Oscillation) and occasional volcanic eruptions. There is, however, evidence for a long-term downward trend in stratospheric ozone, upon which the variations due to these shorter-term influences are superimposed. The longest ground-based series of total column ozone measurements comes from Arosa in Switzerland where data have been collected since 1926. The smoothed data from this site show a slight increase in ozone over the period 1926–73, but a decline of

TABLE 3.2 Ozone Depletion Potentials (Source: WMO, 2003)

Trace gas	Semi-empirical ODP	ODP in Montreal Protocol
CFC-11	1.0	1.0
CF_3Br (Halon-1301)	12.0	10.0
CF_2ClBr (Halon-1211)	6.0	3.0
CCl_4 (carbon tetrachloride)	0.73	1.1
HCFC-22	0.05	0.055
CH_3Br (methyl bromide)	0.38	0.6
HFC-134a	$<1.5 \times 10^{-5}$ (modelled)	

2.9 per cent per decade for the period 1973–97 (Staehelin *et al.*, 1998). The TOMS data for the Arosa area show a similar trend, although the absolute values are slightly different (Fig. 3.8). These data seem to indicate a slight increase since the late 1990s.

Global TOMS data (60°N to 60°S) show an overall decline from about 290 to 300 DU in the late 1970s to 275 to 285 DU in the early 1990s, a loss of about 5 per cent. The most dramatic losses, however, have been recorded in the high latitudes of both hemispheres. Maximum losses occur at about 40 km and 15 km up. There are no significant trends in the tropics, although decadal variability occurs consistent with the 11-year solar cycle. The TOMS data for the Antarctic show a decline of about 60 per cent in ozone in September and October (the Southern Hemisphere spring). The overall rate of loss in high southern latitudes is about 22 per cent per decade. The fastest decline occurred between 1982 and 1987, but then stabilized until 1991–93. The Antarctic ozone holes of 1992 and 1993 covered large areas and were the deepest recorded up to that date. In October of both

these years >99 per cent of O_3 was depleted between 14 and 19km in the atmosphere. This massive loss of ozone has been attributed to the effects of enhanced aerosol loading from the Mount Pinatubo eruption. Annual average global total column ozone reached its lowest levels in these years (WMO, 2003). Antarctic ozone holes in the 1990s were longer lasting than in previous decades. In 1998, minimum ozone south of 40°S, did not get above 160 DU until early December. The largest area ever covered by an Antarctic ozone hole was recorded in September 2000 (29.78 million square kilometres), with the hole of September 2003 coming close (28.7 million square kilometres) (Fig. 3.9).

The 2003 minimum was 103 DU. In contrast, the hole of 2002 had been quite small, due to different meteorological conditions less conducive to ozone depletion.

The effects of cold conditions have also been seen in the Arctic, where conditions tend to be more variable than over the Antarctic. The winter of 1991–92 saw the development of a persistent anticyclone in the troposphere over the

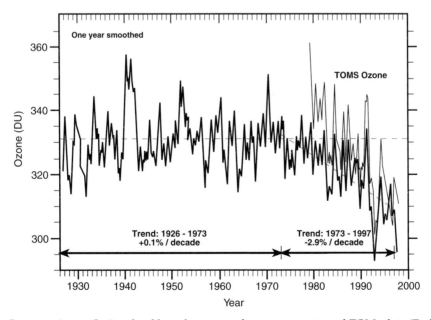

FIGURE 3.8 Ozone at Arosa, Switzerland based on ground measurements and TOMs data (Redrawn from Fig. 9.01b in *Stratospheric Ozone* an electronic textbook. Funded by NASA and available at http://www.ccpo.odv.ed/SEES/ozone/oz_class.htm)

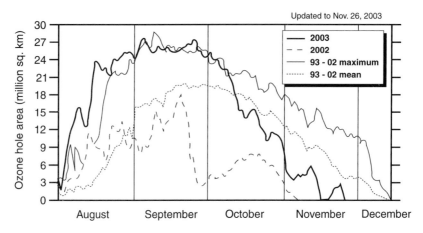

FIGURE 3.9 Area of the Antarctic ozone hole based on NOAA SBUV/2. 2003 compared with past ten years (Redrawn from http://www.cpc.ncep.noaa.gov/products/stratosphere).

Atlantic/north-west Europe, which raised the tropopause and reduced the O_3 column. The Arctic lower stratosphere became very cold, PSCs were formed and there was major O_3 loss. A very cold spring in 1996 had the same effect and resulted in an ozone hole extending as far as the UK. Arctic ozone loss over the period January to March is usually about 20–25 per cent, but the very cold spring of 1996/97 resulted in about 50 per cent ozone loss near 20 km altitude (SORG, 1999). Severe winters in 1999–2000 and 2002–03 also led to major losses. The loss of ozone in Northern Hemisphere mid-latitudes (30° to 60°N) averaged 6 per cent per decade between 1979 and 1994 in winter and spring, with a faster rate of decline in the 1980s than in the 1970s. This accelerating trend continued through to the mid-1990s, when the winter/spring loss rate increased to 8 per cent per decade. Total column ozone showed some increase in the late 1990s (WMO, 2003).

The policy response, at the global scale, to the depletion of stratospheric ozone was remarkably swift. In 1987, the UN Montreal Protocol on Substances that Deplete the Ozone Layer was agreed. This underwent further revisions in 1995, 1997 and 1999. 2004 saw the 16th Meeting of the Parties in the Czech Republic. The Montreal Protocol and its terms are discussed in more detail in Chapter 7. It is clear that this protocol has been successful in first slowing the increase in ozone depleting substances and now in reducing their abundance. The trend in the troposphere is already down and the loading in the stratosphere apparently peaked in the 1990s (WMO, 2003). The rate of decline in stratospheric ozone in mid-latitudes has also slowed. As active chlorine and bromine will only be removed slowly from the stratosphere, it is clear that significant ozone depletion in high latitudes will continue to occur for at least the next 10 to 20 years. The latest WMO estimates (WMO, 2003) suggest that substantial recovery in ozone should occur by the mid-twenty-first century if signatories to the Montreal Protocol and its Adjustments meet their emissions reduction obligations. The WMO continue to forecast nearly complete destruction of ozone in the Antarctic spring, but conditions over the Arctic in its spring season will be much more dependent on climatic variability – how cold it is – and therefore more difficult to forecast. Another unpredictable factor will be the occurrence of major volcanic eruptions. Trends in ozone depleting substances and in the occurrence and size of ozone holes are subject to periodic scientific review – the publications are available through the UNEP website http://www.unep.org/ozone/sap2002.shtml.

3.3.2 The effects of stratospheric ozone depletion

A lot of interest has focused on the increase in UV-B radiation reaching the earth's surface as ozone is lost from the atmosphere. Ozone amount is not the only factor, however, with cloudiness, aerosols and surface albedo also playing a part (WMO, 2003). High quality measurements of ground level UV-B have only been available since the late 1980s. A range of factors will affect the amount of UV-B reaching the surface at any given point (e.g. height of the sun above the horizon, cloudiness), but it is clear that ozone loss has been matched by an increase in UV-B (Fig. 3.10).

Ground level measurements are becoming more reliable and more numerous, but still have to account for a great deal of local variability which can be eliminated from satellite measurements. High and mid-latitude sunburning (erythemal) UV irradiance is estimated to have increased by 6–14 per cent over the last 20 years (WMO, 2003). Long-term measurement data from Toronto, Canada show an increase of 1.5 per cent per year in UV radiation

at 300 nm between 1989 and 1997 (for the period between May and August). These and other data show anomalous UV peaks associated with very low stratospheric ozone after the Pinatubo eruption.

Satellite data (TOMS) have provided long-term UV estimates for the period 1979–92. TOMS does not measure UV, but collects continuous data on ozone and aerosol concentrations and cloud reflectivity which are used to estimate UV in the erythemal range (290–400 nm). TOMS estimates of surface UV are systematically higher than ground-based measurements at many sites and the satellite estimates seem to be worst in the polluted Northern Hemisphere. Because of a break in the TOMS coverage, long-term trend data are currently only calculated through to 1992. The increase has been 3.7 per cent per decade at 50°–65°N, 3 per cent per decade at 35°–50°N and 9 per cent per decade at 50°–65°S (WMO, 1999). The greatest increases occur in the late winter/spring. As might be expected, there has been little change in low latitudes. Follow on TOMS (Meteor-3, ADEOS) have extended the measurement record, but there is still a gap of 18 months in 1995–96.

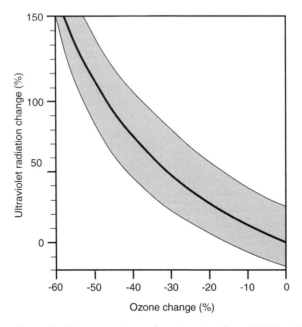

FIGURE 3.10 Envelope of recorded increases in sunburning (erythemal) UV radiation due to ozone decrease (Redrawn from p. xxxvi of WMO (1998) with their permission)

Research on the effects of increased UV-B radiation was initially at quite a low level, but developed considerably during the 1990s. As well as assessing the evidence for effects in a range of organisms and materials, there is also an interest in how to mitigate these effects in the short term (UNEP, 1994). The strongest evidence of a causal relationship between UV-B exposure and human health is skin cancer (malignant melanoma and non-malanoma forms). Damage to the skin can also cause premature ageing, severe photo-allergies, reduced immunological defences and also make vaccines less effective (Goettsch et al., 1998). Eye damage has also been identified, particularly an increasing risk of cataracts associated with cumulative exposure http://www.who.int/uv/health/en/). It is believed that incidences of skin cancers (all types) and cataracts associated with enhanced ozone depletion will peak around the middle part of the twenty-first century (SORG, 1999) and then decline. For humans, it is clear that behavioural patterns are key – overall UV exposure is clearly linked to outdoor activities and holidays in the sun (see WHO, 2003). The effects of UV-B on animals are generally similar to those in humans, with length of exposure playing a key role.

Assessing the effects of increased UV-B exposure on ecosystems is complex as there are a range of interactions with other factors such as exposure to pollutants and the effects of climatic change. Primary producers in aquatic ecosystems (phytoplankton) seem to be most at risk and this would then have knock-on effects for the rest of the food web (e.g. Xenopoulos et al., 2000). It now seems unlikely that increased UV-B exposure will have significant direct effects on terrestrial plants (particularly crops which have been the focus of research). It is possible, however, that chemical and genetic changes caused by UV-B may affect plant development in the long term and the relationships between plants and other organisms such as fungi, decomposing microbes and herbivores (SORG, 1999). (These relationships are not well established). The effects of UV exposure on both aquatic and terrestrial ecosystems could have important impacts on global biogeochemical cycles, altering sources and sinks of greenhouse and other trace gases (CO, CO_2, DMS) (Zepp et al., 1995). These feedbacks need to be considered when looking at the future behaviour of the global climate system.

The second area of concern is the relationship between ozone depletion and climatic change through changes in radiative forcing. As noted above, many CFCs and their substitutes cause positive radiative forcing, but the results of changes in ozone are complex, with both direct and indirect effects. The direct effect of loss of stratospheric ozone is a negative forcing (i.e. a net cooling). The recent IPCC report (IPCC, 2001) estimated this negative forcing as -0.15 ± 0.1 Wm^{-2} since the late 1970s. Earlier estimates (IPCC, 1996; WMO, 1999) ranged from -0.1 to -0.2 Wm^{-2}. All of these estimates are highly uncertain. Ozone in the troposphere gives rise to a positive forcing (warming); the latest estimate of this is $+0.35 \pm 0.15$ Wm^{-2} (IPCC, 2001). The reliability of the estimates of forcing from ozone in both the troposphere and the stratosphere is believed to have improved since IPCC, 1996.

The indirect effects of ozone occur through changes in the availability of the oxidant OH and changes in tropospheric ozone as a result of changes in the stratosphere. A decrease in O_3 results in increased UV-B penetration, which in turn results in an increased concentration of OH in the troposphere. Given its importance as an oxidant of many greenhouse gases (such as CH_4, HCFCs and HFCs), this increase in OH would decrease the atmospheric lifetimes of these gases and hence effectively have a negative radiative forcing. It is also possible that concentrations of another oxidant, hydrogen peroxide (H_2O_2), also increase with increasing UV-B. Although changes in surface ozone concentrations are highly variable, it has been noted that large stratospheric ozone losses also result in losses at the surface. Surface O_3 at the South Pole declined by 17 per cent between 1976 and 1990. The latest estimate is that this change in low level ozone, driven by depletion in the stratosphere, results in a negative forcing. WMO (1999) report a negative forcing from indirect ozone effects of -0.03 to -0.04 Wm^{-2}.

The sensitivity of the global climate system (specifically surface temperature) to different radiative forcings, is of key interest to both scientists and policy makers. The latest WMO report (2003) suggests that the climate may be slightly more sensitive to forcings from stratospheric ozone change than it is to an equivalent forcing from greenhouse gases such as CO_2.

3.4 Atmospheric and climatic change

Arrenhius' interest in explaining past climatic changes through variations in the CO_2 content of the atmosphere has, to a great extent, been replaced by a desire to determine whether changes in the composition of the atmosphere, brought about as a result of human activities, are changing the global climate. IPCC has taken a lead role in assessing the evidence. It is clear that global temperature changes are in phase with concentrations of the recognized greenhouse gases (see the ice core records). As described above, however, there are a host of complex interactions to consider. Understanding what determines the atmospheric lifetimes of important components of the atmosphere and what may affect the capacity of global sinks, for C for example, is key. Whatever the understanding of the processes and the legislation brought in to control emis-

sions, there will always be significant year to year variability as a result of the inherent variability of the climate system and apparently random factors such as the occurrence of major volcanic eruptions.

While early studies of radiative forcing of climate looked only at CO_2, now a whole range of gases and aerosols are considered, together with factors such as the direct and indirect effects of stratospheric ozone depletion and feedbacks resulting from changes in aerosols, surface hydrology, vegetation change and surface albedo (Shine and Forster, 1999). The most recent IPCC estimates (2001) of globally averaged radiative forcing reflect this increased understanding (Fig. 3.11).

As understanding of the complexities increases, so the global climate modellers must attempt to represent these processes in their models (see Chapter 6).

Increasing concentrations of greenhouse gases over the last two centuries have been matched by an overall increase in global surface temperatures, particularly during the twentieth century. Global average surface temperature has increased by 0.6 ± 0.2 °C since the late nineteenth century (IPCC, 2001). The combined land and sea temperature record shows that this increase has not been steady (Fig. 3.12) with an early peak in the 1940s followed by cooling and

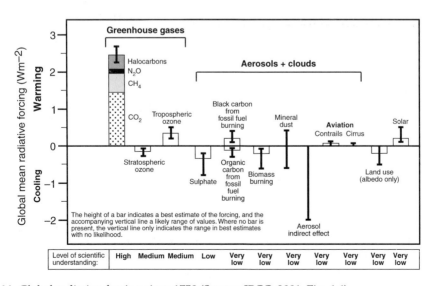

FIGURE 3.11 Global radiative forcing since 1750 (Source: IPCC, 2001, Fig. 6.6)

then another rise through much of the 1980s and 1990s and into the early twenty-first century. 1998 and 2002 were the warmest years of the whole series, with 1998 topping the list at 0.58°C above the 1961–90 mean (Palutikof, 2003).

There is a need for caution when viewing such graphs. The instrumental climatic record is short and in many parts of the world sparse and/or unreliable. The nineteenth century saw the end of the so-called 'Little Ice Age' which tends to enhance the appearance of warming since that time. Where historical and other proxy data are used to extend temperature records, the picture until the 1980s can look quite different. The importance of the longer view is being recognised by the inclusion of a chapter on palaeo climates in the IPCC 4th Assessment Report, due in 2007. A detailed look at year to year variations shows the cooling effects of major eruptions such as Mount Pinatubo and the influence of ENSO. Warm El Niño phases, such as that in 1997–98, enhance warming trends, while cool La Ninas (1999) lower overall temperatures.

In spite of all the variability of climate system and all the uncertainties in understanding, IPCC (1996) decided that 'the balance of evidence suggests a discernible human influence on global climate'. This tentative finding was put more strongly in the IPCC's Third Assessment (2001), which stated that:

- The warming over the past 100 years is very unlikely to have been due to internal variability.
- Reconstructions of climate data for the past 1000 years also indicate that this warming was unusual and unlikely to be entirely natural in origin.
- Detection and attribution studies consistently find evidence for an anthropogenic signal in the climate record of the last 50 years.

While a significant body of scientific opinion now believes that global atmospheric pollution is altering our climate, it is still not universally accepted. Developing successful mitigation strategies will demand great care. Reducing the ozone hole may be desirable, but may enhance global warming; cuts in SO_2 emissions may reduce acid deposition (Chapter 4), but will reduce an element of negative radiative forcing; planting more trees may – or may not – add a sufficient carbon sink to allow some countries to argue against cuts in their emissions. Some of the policy implications of this are explored further in Chapter 7.

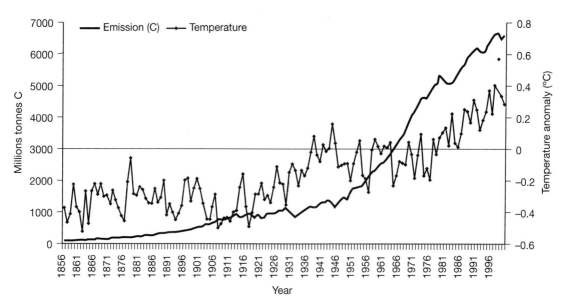

FIGURE 3.12 Global surface temperature anomalies and their relation to CO_2 emissions (CO_2 emissions from Marland *et al.*, 2003, global annual temperature anomalies from Jones *et al.*, 2001)

Regional scale air pollution

4.0 Introduction

This chapter addresses three issues which are primarily felt to be of concern at the regional level (i.e. operating over spatial scales of 100s to 1000s km): acid deposition, heavy metals and tropospheric ozone. The first two of these represent pollutants whose major anthropogenic source is fossil fuel combustion, while ozone is a product of photochemistry. The differing origins of these pollutants is particularly significant in terms of our potential ability to control their emission or production and hence to reduce any environmental impacts. Most of the pollutants discussed here are also important at the global scale (see Chapter 3).

4.1 Acid deposition

Although still commonly referred to as 'acid rain', it has been known for some time that a variety of processes can result in the transfer of acidifying pollutants from the atmosphere to the surface. The major pollutants of interest are compounds of sulphur (particularly sulphuric acid (H_2SO_4)) and oxidized and reduced nitrogen (particularly nitric acid (HNO_3), nitrate (NO_3) and ammonium (NH_4)). Hydrogen chloride can also be a large source of acidity, but its effects tend to be more localized than the others because it is rapidly removed from the atmos-

phere. These pollutants may be deposited in gas, aerosol or liquid phase and will have varying atmospheric lifetimes and hence transport distances. The term 'acid rain' was first popularized by Robert Smith, the UK's first Alkali Inspector, in 1872 following his work on rainfall composition around Manchester.

Concerns over widespread ecosystem damage as a result of acid deposition first came to light in Europe in the late 1950s, in relation to freshwaters in southern Sweden, and as a possible cause of forest damage in the then West Germany in the 1970s. By the late 1970s, early air pollution modelling studies and the analysis of back trajectories of polluted air masses, carried out under the remit of an OECD study of the Long Range Transport of Air Pollutants, had established that a significant element of acid deposition was a transboundary (i.e. long-range) air pollution problem. It was shown, for example, that air masses reaching Scandinavia started out about two days earlier over central Europe and the UK where SO_2 emissions were very high. As a result of this long-range transport, policy initiatives to reduce emissions and ecosystem damage have had to be implemented at the transnational level.

4.1.1 Emissions of acidifying pollutants

Across Europe and North America, signatories to the 1979 UNECE Convention on Long Range

Transboundary Air Pollution (LRTAP) (see Chapter 7) are obliged to provide estimates of their annual, or in some cases monthly, emissions to air. These emissions data are collected in a consistent way across the member states and produced at a spatial scale of 50 km × 50 km. The structures for collecting emissions data have been described in Chapter 2. EMEP publish emissions data for European countries, a range of sea areas (e.g. the North Sea, the Baltic, the Mediterranean) and other source types such as volcanoes. Emissions data for Canada and the USA are also held as these countries are signatories to the LRTAP Convention. The emissions data are required to:

- assess the state and trends of air pollution in Europe;
- establish the compliance of the parties with protocol commitments; and
- provide a basis for the development of cost-effective abatement strategies.

The data are made available in annual EMEP reports (e.g. Vestreng and Klein, 2002) and via the EMEP website which is home for the UNECE/EMEP emissions databases through WebDab (http://www.webdab.emep.int). Data for Canada and the United States are not listed in the printed EMEP reports.

4.1.1.1 Europe

Many European countries compile their own detailed emissions inventories, and these data are then passed on to supranational organizations such as EMEP (see above). The European Environment Agency (EEA) data services provide an umbrella organization for the compilation and dissemination of emissions inventories for the members of the EU, EFTA countries and a number of eastern European and other states. Emissions estimates are provided for SO_2, NO_x as NO_2 and NH_3 as well as a range of other pollutants, across a range of source categories (http://www.eea.eu.int – follow Air theme). Earlier Europe specific inventories compiled by CORINAIR (CORe INventory of AIR emissions methodology) are also available (e.g. for 1990 and 1994). The

CORINAIR data show emissions of pollutants both in tonnes per year and as tonnes/1000 population. Expressing emissions on the basis of population sometimes has surprising results. For example, in 1994 Iceland had the highest emissions of NO_x per 1000 population, Italy the highest SO_2 (largely because of volcanic activity) and Greece the highest emissions of NH_3.

Emissions details for a number of eastern European countries are available through their State of the Environment Reports. In the Czech Republic (now in the EU), for example, SO_2 emissions in 2001 were a staggering 86 per cent lower than in 1990 (http://www.nfp-cz.eionet.eu.int). Reports from some countries are less up to date than the Czech example. Poland (also a new EU member) emitted 4 million tonnes of SO_2 in the late 1980s, about 10 per cent of the European total. Between 20 and 25 per cent of Poland's SO_2, NO_x and dust emissions came from the Katowice region. The political changes of 1989 brought about major economic restructuring and more investment in abatement technology and, as a result, SO_2 emissions in 1995 were 40 per cent lower than they were in 1989. Between 1989 and 1991, Poland's industrial production fell by more than 35 per cent and this was matched by a decline in emissions of air pollutants. Since 1992, however, industrial production has increased while SO_2 and other emissions have continued to fall.

4.1.1.2 North America

For the USA and Canada more detailed information concerning emission sources and trends through time are available from the US Environmental Protection Agency (EPA) and Environment Canada (EC) respectively. Trends in US air pollution, including emissions, are summarized in annual reports which are available over the Internet (http://www.epa.gov/airtrends). Emissions of SO_2 in the USA are dominated by the power supply industry (> 65 per cent in 2003), particularly coal fired power stations. NO_x emissions come mainly from vehicles (49 per cent) followed by power generation (27 per cent). Emissions data for

Canada are available at http://www.ec.gc.ca/pdb/ape/cape_home_e.cfm by sector (e.g. power generation, transportation) and by province. In contrast to the USA, industrial processes are the largest source of SO_2 in Canada (61 per cent), with power generation contributing only 21 per cent. The source allo-cation for NO_x is more similar between the two countries with vehicles contributing 59 per cent of Canada's total NO_x emission.

SO_2, NO_x and NH_3 emissions in selected UNECE countries are shown in Table 4.1. Information on NH_3 emissions is sparse and generally less reliable than for SO_2 and NO_x.

TABLE 4.1 UNECE emissions in 1000s tonnes/yr^{-1} for selected countries (Data from Vestreng, 2003 and US EPA Technology Transfer Network)

SO_2	1980	1985	1990	1995	2000
Belgium	828	400	362	257	165
Czech Republic	2257	2277	1881	1089	264
Germany	7514	7732	5322	1939	638
Italy	3757	1901	1651	1322	758
Norway	137	98	52	33	27
Poland	4100	4300	3210	2376	1511
Canada	4643	3692	3210	2633	1211
USA	23501	21074	20989	16830	14803*
NO_x (as NO_2)					
Belgium	442	325	334	359	329
Czech Republic	937	831	544	368	321
Germany	3334	3276	2728	1984	1584
Italy	1638	1614	1938	1768	1371
Norway	91	213	224	221	224
Poland	1229	1500	1280	1120	838
Canada	1959	2038	2104	2032	2011
USA	22501	21308	21584	21714	21046*
NH_3					
Belgium	89@	89	99	100	81
Czech Republic	156@	156@	156@	86	74
Germany	835	857	736	603	596
Italy	479	487	466	461	437
Norway	23	23	23	26	25
Poland	550	550	508	380	322
Canada	ND	ND	ND	540	ND
USA	ND	1685$	4731$	ND	4523*

* Figure not officially submitted to UNECE (from US EPA)
@ Expert estimate
$ limited information on sources included

4.1.1.3 Asia

Estimates of emissions across Asia have been developed more recently than those for Europe and North America. Kato and Akimoto (1992) present emissions totals for SO_2 and NO_x from 25 Asian countries east of Afghanistan and Pakistan for the years 1975, 1980, 1985, 1986 and 1987. Across the region as a whole, SO_2 emissions increased by 3.8 per cent per year and NO_x emissions by 4.1 per cent per year over the period of the study. Only Japan was implementing significant emissions controls (such as flue gas de-sulphurisation on power stations and controls on vehicles). China was the largest single emitter of both SO_2 and NO_x, contributing 68.6 per cent and 47 per cent of the total respectively (1987 figures). China, India, South Korea, Thailand and Taiwan emitted 92 per cent of the region's SO_2 and these countries, plus Indonesia and North Korea, emitted 92 per cent of the NO_x.

In the early 1990s, it was thought likely that there would be a very rapid increase in SO_2 emissions from Asia as a result of economic growth. This has proven not to be the case (Streets et al., 2001) as a result of changes in fuel type, environmental controls and economic slow-down. High projected SO_2 emissions estimates were used in modelling work through RAINS-ASIA (see Chapter 6) and this should be taken into account when the results from that project are reviewed. NO_x emissions, however, continue to rise with an increasing proportion coming from vehicles (particularly in China) (Streets et al., 2003).

4.1.1.4 General trends in emissions

Long-term (historical) emissions estimates have only recently become available. Estimates of global anthropogenic emissions (on a $1° \times 1°$ grid) have been compiled for the period 1890–1990 for a range of gases by Van Aardenne et al. (2001) and are available through the EDGAR-HYDE website hosted at RIVM (http://www.arch.rivm.nl/en/). SO_2 emissions have been compiled for Europe for the period since 1880 by Mylona (1996). Trends in emissions for Europe (excluding the former USSR and Turkey) are shown in Fig. 4.1.

Emissions rose steadily through the first half of the twentieth century and were dominated by emissions from coal combustion until after World War II. Total emissions increased sharply from about 1950 until 1980, with a marked increase in the contribution from liquid fuels (oils).

Emissions of SO_2 have shown major reductions across Europe since 1980, declining by about 55 per cent. This has partly been in response to legislation (see Chapter 7) and partly due to changes in the patterns of economic activity. Political and economic changes

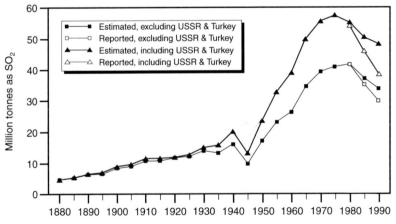

FIGURE 4.1 Total annual sulphur dioxide emissions across Europe since 1880 (Source: Mylona, 1996, Fig. 3. Reproduced with the permission of the Norwegian Meteorological Institute/EMEP/MSC-W)

in eastern Europe since 1990 have had a particularly pronounced impact on emissions from industrial sources in these areas. Canada has shown a similar pattern. In the USA, however, the general decline in emissions from 1980 onwards was reversed between 1995 and 1997, but has subsequently resumed a downward trend (EPA, 2003a). Across Asia, rapid economic development, especially in south-east Asia, initially drove up SO_2 emissions. Industrial expansion in China was largely based on coal burning, with coal consumption increasing by an order of magnitude between 1953 and 1982. A 'business-as-usual' scenario (Whelpdale and Kaiser, 1996) estimated that SO_2 emissions in China would increase from 20 million tonnes/yr^{-1} in 1987 to 50 million tonnes/yr^{-1} by 2020. As described above, these early estimates now seem unduly pessimistic with emissions in 2000 being about the same as in 1987. Emissions trends for some Asian countries are illustrated in Fig. 4.2. As the data come from two different sources, the totals for the two time periods represented in the figure are not directly comparable.

NO_x emissions show a close relationship to levels of car ownership. Across Europe, emissions peaked in the mid- to late 1980s and then started to decline. Since the overall peak in Europe in 1990, NO_x emissions have declined by about 20 per cent, although there is considerable variability between the different countries. In the USA and Canada, emissions of NO_x were reduced in the early 1980s, but then increased again. The introduction of catalytic convertors has significantly reduced the emission per vehicle, but there is continued growth in the vehicle fleet. Overall, NO_x emissions in the USA have decreased 12 per cent since 1993.

Across Asia, emissions of NO_x have generally increased and seem set to continue to do so.

4.1.1.5 Emissions and trends in the UK

In the UK, a National Atmospheric Emission Inventory is compiled for the government to meet a number of national and international requirements (see Chapter 2). All the data compiled by the NAEI are available on the Internet (http://www.naei.org.uk) and show totals, the contribution by different source categories and the spatial distribution of the emissions (on a 1 km × 1 km grid). Since 1988, emissions of SO_2 and NO_2 from large combustion plant (> 50 MW capacity) have been made available by the UK Environment Agency on a plant by plant basis; more recently this has been extended to some emissions from processes covered by the Chemical Release Inventory. Emissions from other types of sources are calculated by applying an emission factor based on a limited number of actual measurements. Estimates of ammonia emissions are highly uncertain (RGAR, 1997), but the main sources are farm animals (especially cattle) and their waste. Trends in UK emissions are summarized in Table 4.2.

UK emissions of SO_2 are still dominated by electricity generation (coal and oil fired power stations) which contributed 66 per cent of the total in 2001. Other industrial combustion (13 per cent) was the next largest source. Domestic sources now make only a small contribution except in areas such as Belfast. There has been a decrease of > 83 per cent in the UK's emissions of SO_2 since 1970 which showed the highest emissions since estimates began. This has been due to changes in industrial activity (particularly a decline in heavy industry), switches in

TABLE 4.2 UK emissions tonnes 1000s/yr^{-1} (Dore *et al.*, 2003)

	1970	1980	1990	1995	2000
SO_2	6460	4854	3719	2365	1188
NO_x (as NO_2)	2501	2581	2759	2174	1737
HCl	337	310	276	161	87
NH_3			341	319	297

the fuel used for electricity generation and the decline of domestic coal burning.

The main source of NO_x (46 per cent) is road transport, followed by power generation. Although emissions have declined since 1970 (by about 30 per cent), the trend was relatively flat through the 1970s to mid-1980s, rising to a peak in 1989, before falling back. The increase was due to emissions from petrol vehicles, especially cars. Since 1989, emissions from the transport sector have declined due to tighter controls on emissions from diesel vehicles and the introduction of catalytic convertors on petrol cars. There has also been a substantial reduction (51 per cent) in emissions from power stations.

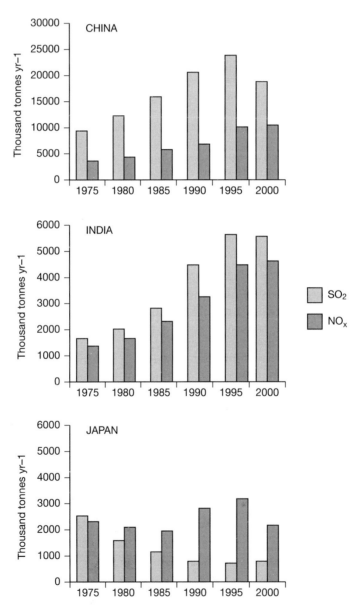

FIGURE 4.2 Estimates of SO_2 and NO_x emissions from China, India and Japan (Based on data in Kato and Akimoto (1992) for 1970, 1980 and 1985, and Streets *et al.* (2003) for 1990, 1995 and 2000)

4.1.2 Transformation and deposition of acidifying pollutants

By the 1980s, it was clear that there was a need for a much greater understanding of the processes of transformation and deposition affecting potentially acidifying compounds. The initial focus of this work was sulphur, but more recently attention has turned to nitrogen which is considerably more complex. Primary pollutants, such as SO_2 and ammonia (NH_3), undergo chemical transformation as they are dispersed in the atmosphere and produce secondary pollutants (e.g. sulphate (SO_4) and nitrate (NO_3)). These secondary pollutants often display different properties from their parent, primary pollutants (e.g. being in aerosol form, rather than as a gas) and this affects the way they behave in, and are removed from, the atmosphere. The chemistry of the atmosphere is discussed quite fully in a range of textbooks (e.g. Brimblecombe, 1996; Seinfeld and Pandis, 1998) and only a summary will be provided here. Current understanding of the behaviour of acidifying pollutants in the UK is described in RGAR (1997) and NEGTAP (2001).

The most important transformations are brought about by oxidation. There are a range of oxidants and oxidation may take place in the gas or aqueous phase. The primary oxidant in the troposphere is OH (the hydroxyl radical) which results in photochemical conversion. Ozone (O_3) and hydrogen peroxide (H_2O_2) can also be important. The major oxidation pathways for acidifying pollutants have been identified as set out below.

Sulphur
Photochemical oxidation of SO_2 gas to soluble sulphate aerosol:

$$SO_2 + OH\ (+O_2 + H_2O) \rightarrow H_2SO_4 + HO_2$$

If the SO_2 dissolves in water droplets in cloud, bisulphite (HSO_3^-) or sulphite ions are formed, depending upon the pH. These may then be oxidized by either O_3 or H_2O_2 to yield sulphate in solution:

$$HSO_3^- + O_3 \text{ OR } H_2O_2 \rightarrow SO4_2^- + H^+$$
$$SO_3^{2-} + O_3 \text{ or } H_2O_2 \rightarrow SO_4^{2-}$$

H_2O_2 is rapidly consumed by this reaction so the importance of this reaction is limited by the availability of H_2O_2 in the atmosphere. Oxidation by O_3, however, is really limited only by pH, becoming very slow when pH drops below 4.5. Because of this pH dependence, the importance of O_3 oxidation of sulphur is dependent on the presence in the atmosphere of neutralizing species such as ammonia.

Oxidized nitrogen
Gaseous nitrogen oxides play a role in acidification, eutrophication and ozone production (see below). Many of the reactions are reversible. Oxidation of NO by O_3 to yield nitrogen dioxide:

$$NO + O_3 \rightarrow NO_2 + O_2$$

During daylight hours, however, NO_2 is rapidly converted back into NO as a result of photolysis (sunlight). This rapid interconversion (over time periods of minutes) is described as a photostationary state. The mechanism also produces O_3. NO can also be converted to NO_2 by reactions with free radicals; the photolysis of the resulting NO_2 is a net source of O_3.

NO_2 is also oxidized by O_3 to form the nitrate radical, but in daylight the NO_3 is rapidly photolysed. At night, the NO_3 persists and reacts further with NO_2 to give dinitrogen pentoxide (N_2O_5). This then reacts with water vapour to form nitric acid.

The key reaction for removing NO_x from the atmosphere during daylight is the oxidation of NO2 by the hydroxyl radical to produce gaseous nitric acid (HNO_3). Nitric acid is readily deposited, dissolved in water droplets or reacts with ammonia to form particles:

$$NO_2 + OH \rightarrow HNO_3$$

OH can also oxidize NO to yield nitrous acid (HONO). HONO is rapidly broken down by sunlight (back into OH and NO), but can be dissolved into water vapour producing nitrite ions. In this aqueous phase, HONO (or NO_2) can be oxidized by H_2O_2 to produce nitrate.

Ammonia
Ammonia gas is highly reactive and can be readily deposited back to the surface close to its place of origin. Ammonia also reacts with

acidic compounds produced by the oxidation pathways described above (H_2SO_4, HNO_3, HCl) to yield particulate ammonium which can then be removed from the atmosphere in rain.

4.1.2.1 Deposition

The direct deposition of gases and particles onto a surface is known as dry deposition. For acidification, the gases of interest are SO_2, NO_2, HNO_3 and HCl. NH_3 is also a contributor, because once oxidized to NH_4 (ammonium) it can release acidity. Although it is possible to measure dry deposition, this is rarely done. The standard procedure is to model dry deposition based on measurements of gas concentration and an estimate of surface resistance which is expressed as a deposition velocity. This resistance depends on factors such as vegetation height, the affinity of a particular pollutant for an individual type of surface and wind speed. In general terms, dry deposition is important close to sources of pollutants, while wet deposition (see below) is more important remote from sources. The current understanding of dry deposition processes in the UK and the way that this has been implemented to provide estimates of dry deposition fluxes is described in NEGTAP (2001). Ammonia is dry deposited very efficiently and it is believed that about 40–50 per cent of deposition of reduced N in the UK comes by this route. SO_2 is also quite readily dry deposited, with the rate of deposition to vegetation being determined by whether or not the stomata are open, and by surface wetness (deposition is more efficient on wet surfaces than dry). In contrast, NO_2 is not readily dry deposited and it is estimated that this pathway contributes only 25 per cent of oxidized N deposition to the UK. Nitric acid gas is, however, very readily dry deposited. A national network for measuring nitric acid and aerosol concentrations across the UK came into operation in 1999. The results from this network have had a significant impact on estimates of dry deposition of oxidized N across the country (see below). Information about the network can be found at http://www.edinburgh.ceh.ac.uk/pollution/home.htm.

The direct deposition of water droplets from clouds is called cloud water interception or occult deposition. This process becomes important in areas where cloud cover regularly blankets the surface (usually on hill tops). The concentrations of pollutants in hill cloud have been found to be significantly higher than those in rain. The typical mean value for NO_3 and SO_4 in the UK shows the ratio between hill cloud and rain to be about 5:1. The deposition of cloud droplets will be enhanced in forested upland areas because of their surface roughness, so inputs of acidity in forested areas regularly under cloud may be particularly important.

Wet deposition occurs due to the removal of pollutants from within clouds or beneath clouds as precipitation falls. It is dominated by the removal of acidic aerosol species and so occurs largely remote from sources, once the chemistry of the air has had time to evolve. Wet deposition, therefore, usually predominates in rural and upland areas. The amount of wet deposition is calculated by measuring the concentration of the pollutants in rain (SO_4, NO_3, NH_4) and then multiplying this by the amount of rainfall at a site. Where rainfall is known to contain ions derived from sea spray, the measured concentrations are corrected for this (especially for SO_4) so that the data reflect non-sea salt sulphate. This correction does not need to be applied for nitrate or ammonium.

4.1.2.2 What goes up, must come down ...

Across Europe, the EMEP network was set up under the terms of the Convention on LRTAP to provide a Europe-wide perspective on pollution (see Chapter 2). These sites usually collect data on a daily basis using wet-only collectors (i.e. they only open when it is raining). In common with other major networks, the sites all have clear siting protocols and quality assurance procedures to ensure the reliability and comparability of data from the different sites. In the UK, acid deposition is one of the non-automatic networks in operation (Chapter 2). In the USA, rainfall composition is measured by the National Atmospheric Deposition Network (NADN) of nearly 200 sites. In Canada, precipitation composition is measured by a variety of federal and provincial networks (see Chapter 2).

The number of acid deposition monitoring sites in Canada peaked in the late 1980s. The results from these, together with data from some US networks, are coordinated at the National Atmospheric Chemistry (NatChem) data base in Ontario. Over the period 1977–94 NatChem has used data from 730 sites over different time periods.

In other areas of the world, monitoring tends to be more restricted. There has, however, been a big expansion in monitoring in south-east Asia, with precipitation composition networks being set up in Japan from 1983 and in China from 1982. The Asian Development Bank funded a network of 45 SO_2 monitoring sites across the region starting in 1993/94. The Acid Deposition Monitoring Network in Asia, started in 2000, is described in Chapter 2.

4.1.2.3 Current deposition patterns

Deposition estimates are rarely based on measurements alone. Many countries do not have monitoring networks and where networks are present they are usually scattered. As a result, deposition fields may be generated by a combination of measured values (usually wet deposition) and modelled values (usually dry deposition) or be entirely modelled. Acid deposition may be expressed in a variety of units including kg S or N ha^{-1} yr^{-1}, mg S or N m^{-2} yr^{-1}, or in equivalents of H$^+$ e.g. eq ha^{-1} yr^{-1}. Equivalents are useful units for making direct comparisons between pollutants and for comparing with critical loads (see below).

Across the wider European region, EMEP produces figures for the deposition of S and oxidized and reduced N, using computer models (see Chapter 6). Gas and precipitation compositions are mapped from the EMEP monitoring networks but are used for model validation purposes (see Tarrason *et al.*, 1998 and http://www.emep.int/index_data.html). Some individual European countries produce their own national deposition maps based on their own monitoring and modelling procedures.

Maps of S and oxidized and reduced N deposition across the UK at a grid scale of 5 × 5 km for the period 1995–97 have been published in NEGTAP (2001). Total S deposition across the

UK was estimated at 270 thousand tonnes yr^{-1}, of which 56 per cent was wet deposited and 44 per cent dry deposited. For oxidized N the total deposition was about 178 tonnes yr^{-1}, with 51 per cent wet deposited and 49 per cent dry deposited. This estimate of the wet/dry split was different from earlier deposition budgets (where wet deposition was more dominant) because the first data from the nitric acid network were available. Total deposition of reduced N was reported as 209 thousand tonnes yr^{-1}. More recent deposition data (1998–2000) are illustrated in Fig. 4.3.

Budgets for total S and oxidized and reduced N are 262, 178 and 157 thousand tonnes yr^{-1} respectively. Wet deposition is usually greatest in the high rainfall areas of the uplands of the western and northern UK, often quite remote from emission sources. In contrast, dry deposition is more important close to source regions. A simple comparison with the deposition budgets for 1995–97 would suggest that continuing emissions reductions have not had a significant impact on deposition, but the methods used to compile the two datasets are not the same and the impact of the change in methodology has yet to be assessed.

All valid data from the Canadian monitoring sites, together with data from the US National Atmospheric Deposition Program/National Trends Network (NADP/NTN) and Clean Air Status and Trends Network (CASTNET), have been used to map deposition across eastern North America (Environment Canada, 1997). This showed that the highest sulphate and nitrate depositions were occurring across the north-east USA/south-east Canada, around the eastern Great Lakes. A comparison of data from 1980 until the early 1990s shows that the area of highest SO_4 deposition (> 20kg ha^{-1} yr^{-1}) reduced slightly, while there was little change in NO_3. Maps of air concentrations of SO_2, particulate SO_4 and total NO_3 were also produced for eastern North America by combining data from different networks, but no deposition maps have been produced for the region as a whole. Depositions of SO_4 and NO_3 for 1995 and 2002 based on data from the NADP alone are compared in Fig. 4.4. Other mapped data are available from http://www.nadp.sws.uiuc.edu/.

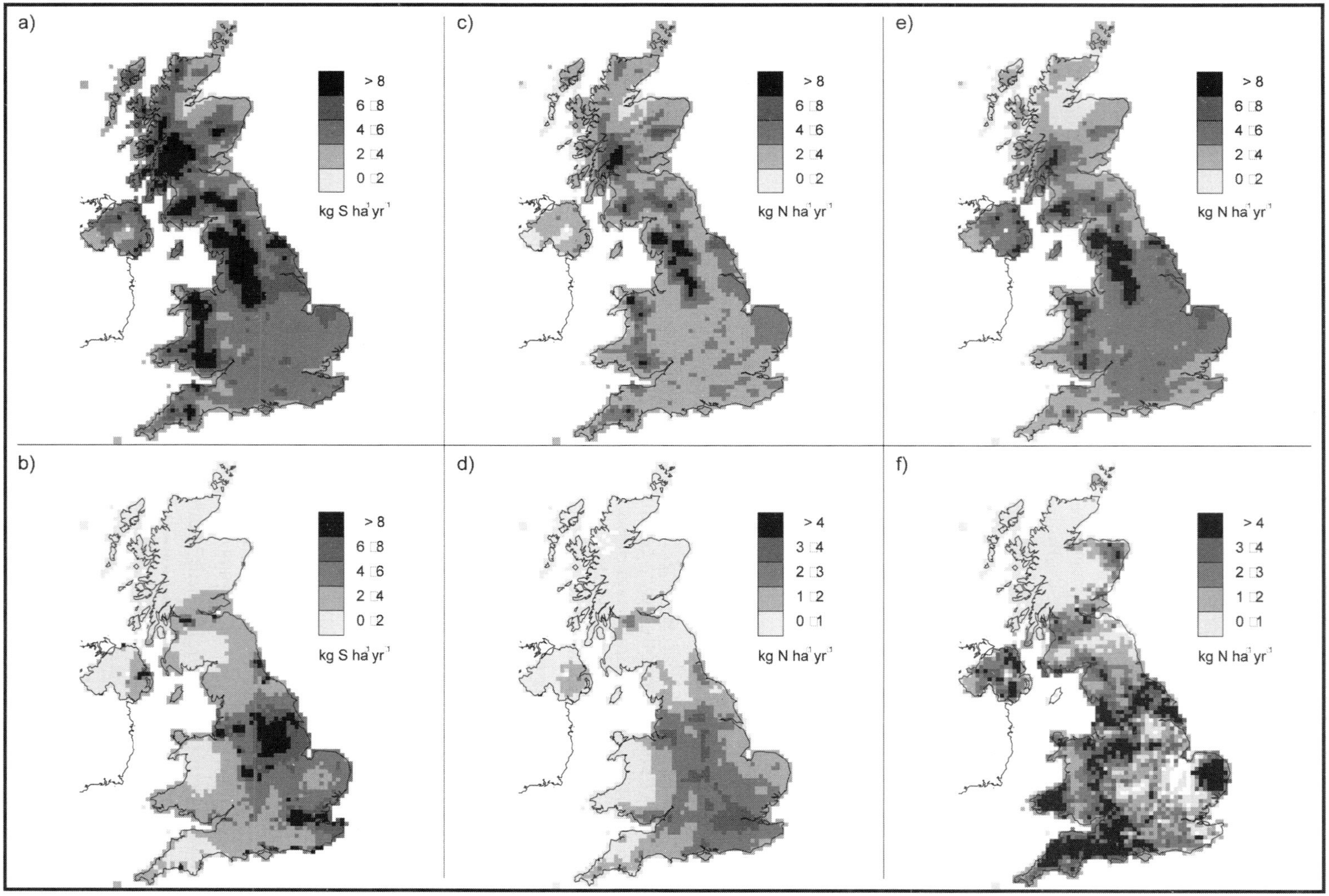

FIGURE 4.3 Wet and dry deposition of S (a and b), oxidized N (c and d) and reduced N (e and f) (in kg) to the UK for 1998–2000 (Data supplied by CEH, Edinburgh)

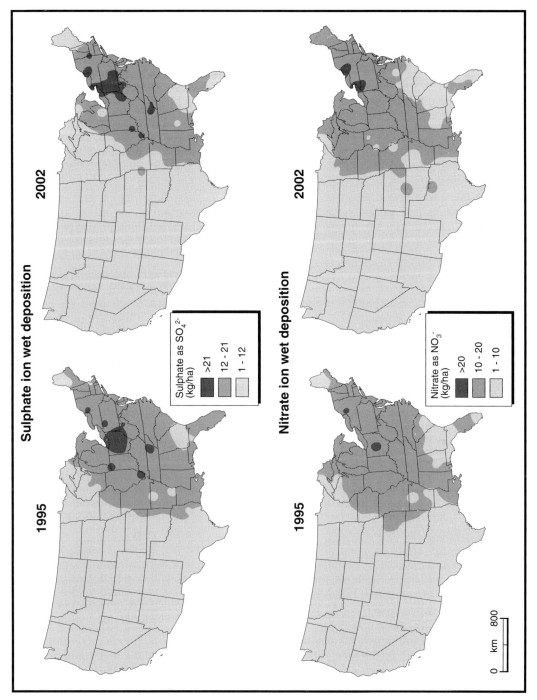

FIGURE 4.4 Annual sulphate and nitrate ion wet deposition (in kg) in the USA for 1995 and 2002 (Based on data from the US National Atmospheric Deposition Program/National Trends Network)

4.1.3 Long-term trends

Direct systematic measurements of pollutants involved in acidification cover only a limited period, although there are a few individual sites with long-term records. Smoke and SO_2 data for the UK go back to the 1960s, although the monitoring sites were concentrated in urban areas (see Chapter 5). A rural SO_2 network was not established until 1991. There was no UK national precipitation composition monitoring network until 1986. The first EMEP monitoring sites came into operation in 1978. It is, however, possible to extend the record by using the information held in ice sheets and lake sediments.

4.1.3.1 Ice cores

Ice cores retain a record of atmospheric composition as gas bubbles and through the ionic composition of the precipitation itself. The sulphate content of ice has a range of possible sources both natural e.g. volcanic and oceanic (including DMS), and anthropogenic e.g. fossil fuel combustion. Early measurements of SO_4 were targeted on reconstructing histories of volcanic activity (Hammer, 1977), but it soon became apparent that the ice cores also provided a record of pollution (Neftel et al., 1985). A study of a core from near Dye 3 in southern Greenland (Mayewski et al., 1990) has yielded a record of SO_4 and NO_3 concentrations extending back to AD 1767. The period 1767–1903 has a sulphate record dominated by biogenic S and large volcanic eruptions. There were significant increases in sulphate concentrations between 1903–60 and 1960–84 which the authors attribute to increasing anthropogenic SO_2 emissions, primarily from the USA. Ice core records from almost all Northern Hemisphere extra-tropical sites (Alps, Arctic, Greenland) show increasing SO_4 concentrations as a result of the long-range transport of pollution. One exception to this is Mount Logan in Canada (5300 m a.s.l.). This site shows no SO_4 trend, apparently because the source of the precipitation is the free troposphere over the North Pacific (Holdsworth and Peake, 1985). It appears that the sulphate content of ice cores from the Antarctic has not been influenced by fossil fuel combustion which has traditionally been concentrated in the mid- to high latitudes of the Northern Hemisphere.

The trends in NO_3 concentrations in Greenland ice also show a pollution effect, particularly since 1940. This increase occurs rather later than that for sulphate, but generally follows the different trends in emissions for these two pollutants (Wolff, 1995). NO_x from anthropogenic sources apparently supplies about 50 per cent of the NO_3 in modern Greenland ice. Similar trends have been reported from other Northern Hemisphere ice cores, although Mount Logan is again an exception. It appears that the pollution signal is seen not only in the total NO_3 concentration, but also in a shift in the seasonal peak from summer (pre-industrial) to late winter/early spring. The ice core from Huascarán, Peru (9°S, 6048 m a.s.l.) records increasing NO_3 levels over the last 200 years, but these are still lower than they were during the mid-Holocene (Thompson et al., 1995). The NO_3 record here appears to reflect emissions from tropical rainforest within the Amazon basin. Ice cores from the Antarctic show no recent increase in NO_3. One record from Vostok (Mayewski and Legrand, 1990) did show an apparent increase, but it now seems likely that part of this may be explained by post-depositional losses of HNO_3 where accumulation rates are low. Perhaps surprisingly, more limited analysis of carbonaceous particles, also produced by combustion, shows no clear changes in concentration between the pre- and post-industrial periods (Cachier, 1995). There has, however, been a change in source type from vegetation burning, to coal and then to oil. The carbonaceous particle record is much clearer in lake sediments (see below).

4.1.3.2 Lake cores

The study of cores of lake sediment provides another long-term record of changes in acidification (Battarbee et al., 1990). Lake sediments generally preserve changes in lake pH, reflected by changes in lake biota, rather than acting as a direct record of changing atmospheric composition. Although declining fish stocks acted as a major trigger for concern over acidification, it was the study of changes in assemblages of diatoms (unicellular algae) through time, that played a key role in identifying the link between the combustion of fossil fuel and declining pH in surface waters.

In the 1980s two major international research programmes were set up to explore acidification in freshwaters: the Surface Water Acidification Programme (SWAP) between Sweden, Norway and the UK and the Paleoecological Investigation of Recent Lake Acidification (PIRLA) in the USA and Canada. When these studies were instigated, there were three possible mechanisms for surface water acidification being considered: fossil fuel combustion, land use change and afforestation using coniferous trees. Diatoms are pH sensitive, with the species living in a lake (or stream) changing as the pH changes. The siliceous valves of diatoms become preserved in lake sediments and can provide a record of pH change through time. The methodology for this has been described extensively elsewhere (Battarbee and Charles, 1987; Birks *et al.*, 1990). By combining diatom records with historical data on land use change, comparing catchments that had and had not been planted with conifers and looking at the carbonaceous particles resulting from combustion preserved in the sediments, it became possible to confirm that the widespread acidification of upland catchments on acid sensitive geologies was the result of acid deposition from fossil fuel combustion. A study of sediment cores from Galloway in south-west Scotland tested the different hypotheses. Diatom-reconstructed pH showed that acidification had occurred by the late nineteenth or early twentieth century. The only hypothesis that could not be rejected was that of acid deposition (Battarbee *et al.*, 1985).

Evidence for the long-range transport of pollutants has also been provided by heavy metals (see below) and spheroidal carbonaceous particles (SCPs). SCPs result from high temperature fossil fuel combustion and it is possible to distinguish SCPs from different fuel types (coal, oil, peat), giving further detail to palaeolimnological studies of acidification in different areas (Rose *et al.*, 1996). Some results of carbonaceous particle and pH reconstruction work are illustrated in Fig. 4.5.

Reconstructions of lake acidification histories have been carried out extensively across Europe and North America and have had a profound effect on the development of policies to reduce the emissions of acidifying pollutants (see Chapter 7). The evidence showed that the blame for acidification in remote areas lay with coal and oil burning power stations. As a result, much of the policy emphasis has been on reducing emissions from these sources.

4.1.4 Damage due to acid deposition

The major foci of concern relating to damage from acid deposition have been freshwaters (and their associated biota), forests and building materials. The loss of fish stocks from lakes in southern Norway was recorded as early as the 1920s. By the 1940s and 1950s similar declines were being seen in southern Sweden. There was a gradual deterioration in conditions between about 1915 and 1950, followed by more rapid change up to 1980. By the late 1980s, 1750 out of 5000 lakes in southern Norway had lost their fish populations, while a further 900 lakes were found to be seriously affected. In southern Sweden, damage to fish stocks was reported in more than 2500 lakes. Monitoring of lake chemistry showed increasingly acid conditions (pH < 5.5), higher concentrations of sulphate and decreasing alkalinity reflecting a loss of acid neutralizing capacity.

Norway's 1000 Lakes survey (Norwegian State Pollution Control Authority, 1987) sampled 1005 (!) lakes in remote areas in the autumn of 1986. The survey showed that 70 per cent of lakes in the south of the country had a pH of less than 5 and zero alkalinity. It also showed high concentrations of labile aluminium in acidified lakes. Aluminium and other heavy metals are usually bound in sediments, but go into solution when pH falls below about 5. The presence of soluble aluminium increases the permeability of fish gills to chloride and hydrogen, something which is usually regulated by Ca^{2+} (low pH systems often also have low Ca) (Wellburn, 1994). The Al can cause clogging of the gills and inhibit calcification of a fish's skeleton meaning that young fish cannot grow to maturity. pH below about 5.5 results in a decline in invertebrate populations, which has a knock-on effect for birds as well as fish. The increased levels of Al and other heavy metals also causes reduced calcification of birds' eggs and the loss of fledgelings.

In the late 1970s, surveys of streams and lochs in south-west Scotland (Galloway) revealed that many surface waters were highly acidic and that some lochs were fishless. Similar conditions were also found in upland Wales. In fact, acidified surface waters occur largely in upland areas of the UK with base-poor geology (low buffering capacity). The UK Acid Waters Monitoring network was started in 1988 to monitor chemical and biological conditions in lakes and streams receiving different acid deposition loadings, at different altitudes

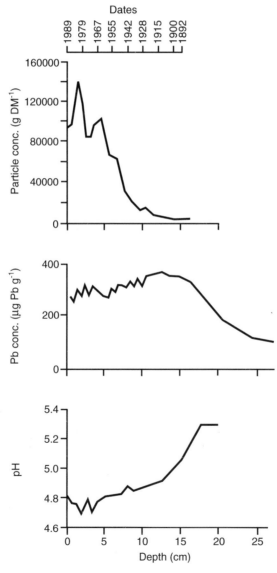

FIGURE 4.5 Reconstructed pH, lead and particle concentration from the Round Loch of Glenhead, Galloway, Scotland (UK Acid Waters Monitoring Network data from Patrick *et al.*, 1995)

and both with and without afforestation. The network originally comprised about 50 sites, but since 1990 has had 22 sites (http://www.ukawmn.ucl.ac.uk). The distribution of sites is now heavily biased towards western and northern Britain.

A review of data from the network's first ten years of monitoring was published in 2000 (Monteith and Evans, 2000). The effects of air pollution on freshwaters were also reviewed by NEGTAP (2001). Lakes in north Wales and Scottish lochs showed declines in fish stocks from the 1950s, while mid-Wales was affected from the 1960s. Fish kills associated with acid 'episodes' (pulses of very low pH runoff associated with storms or snowmelt) seem to be quite rare in the UK. There is, however, evidence for impacts on micro- and macro-invertebrates and hence on organisms for which these are a food supply: falling populations of dippers (*Cinclus cinclus*) have been noted from acidic streams in Wales. It should be noted, though, that even in areas of high atmospheric inputs of S and N not all lakes and small streams will acidify. The response will be strongly mediated by catchment geology and land use. Reductions in emissions of SO_2 since the mid-1980s have resulted in decreased inputs of S to UK catchments. Trends for atmospheric N are less clear. Sulphate concentrations in UK surface waters have decreased in response. Since the mid-1990s, decreases in acidity and labile aluminium have also been recorded. The extent to which this chemical recovery has been matched by biological recovery is discussed further in Chapter 8.

In Canada, acid sensitive environments with significant acid deposition have been identified as occurring in the area south of James Bay (52°N) and east of the Manitoba-Ontario border. There are about 800,000 water bodies in this area and about 8500 of these have been assessed in terms of their chemical and biological characteristics. This sensitive area has the highest numbers of acidified lakes (where acid neutralizing capacity (ANC) <= 0): it has been estimated that there are more than 14,000 acidified lakes greater than 1 ha in size in that area. pH reconstructions, using diatoms and chrysophyte microfossils, have shown that a number

of lakes have acidified over the last 100 to 150 years as a result of acid deposition. The majority of acidification in Canada has been attributed to S deposition (Environment Canada, 1997). Data on the chemical composition of lakes in eastern Canada have been collected since the late 1970s. Biological monitoring has shown impacts on algae, zooplankton, macroinvertebrates, fish, amphibians and birds. A survey of rivers with salmon habitat in Nova Scotia showed that in 14 rivers where the pH of the main river was less than 4.7, salmon were extinct, and in 20 rivers with a mean pH of between 4.7 and 5.0 there was a 90 per cent loss in the salmon population.

Although SO_2 emissions in Canada and adjacent areas of the USA have declined since the early 1980s, the evidence for an improvement in lake status (higher pH or ANC) is patchy. Using a sample of 202 lakes for 1981–94, 51 per cent of lakes showed a decrease in SO_4 concentrations, 48 per cent were stable and 1 per cent showed higher SO_4 levels. It was found that 11 per cent were continuing to acidify.

The relationship between atmospheric pollution, especially acidic deposition, and damage to trees has been more difficult to establish than that for freshwaters and their associated biota. The widespread forest decline that occurred across central Europe from the 1970s was not initially associated with air pollution as the sites were quite remote from major source areas. Although vegetation is affected by high concentrations of SO_2, it is now clear that acidic deposition plays a limited role in affecting tree health. Deposition of N, high ozone concentrations, attack by diseases and pests and adverse climatic conditions are more important. Acid mist can also cause damage (NEGTAP, 2001).

There is evidence for increasing acidity of forest soils. In southern Sweden, for example, the pH of the top horizon of soils beneath beech and spruce forests, fell from about 4.8 in 1927 to about 3.6 in 1982–83. For countries which are highly dependent on commercial forestry, any loss of crop due to pollution damage is of economic significance. One of the real problems, is that trees – and especially spruce trees – increase the rate of S deposition. A study of tree health in the Black Triangle region between

Germany, Poland and the Czech Republic (Cerny and Paces, 1995) showed that there was nearly 43 per cent defoliation of spruce trees. This area lies on acidic granitic and metamorphic bedrock and is close to industries that burnt S-rich coal. It was found that S deposition beneath the forest canopy was more than 70 kg S ha^{-1} yr^{-1} and that the pH was less than 4. Tree health was being further undermined by an outbreak of bark beetle.

In the UK, the Forestry Commission started formal surveys of tree health in 1984 and from 1987 a standard methodology of assessment was applied across Europe. The results from the surveys are available as Information Notes through http://www.forestry.gov.uk. (It is worth noting that the apparent decline in tree health in the UK between 1985 and 1987 may actually have been the result of the change in methodology.) The 1990 survey showed a marked deterioration in the condition of broadleaved trees, based on crown density loss. However, although trees in the UK appear to be in a poor state of health compared with other European countries (DOE, 1993), the link to pollution has still not been clearly established. Climatic stress (drought, frost), nutrient deficiencies, insect attack and fungal disease can all cause damage. Experimental data are mainly short term and come from young trees. It seems unlikely that rural SO_2 or NO_2 concentrations are high enough to retard growth; acid mist will, however, probably slow growth. A more recent Forestry Commission programme 'Environmental Monitoring and Evaluation of Forest Ecosystems', will also be making a contribution to the assessment of impacts of S and N deposition.

Extensive damage to buildings has been recorded across Europe from the Parthenon in Athens to Cologne cathedral. A marked deterioration in the condition of medieval stained glass has also been recorded over the last 30 years. Evidence for the effects of pollutants comes from two sources: examination of buildings using photographs, historical records etc and from the experimental exposure of different types of materials at sites where pollution concentration data are also collected. The main pollutants of concern are SO_2, NO_x, CO_2 and particulates. Dry deposition of SO_2 may be the most important and calcareous materials such as limestone, marble and some sandstones are most at risk. Iron, steel, zinc and low-silica glass are also vulnerable. There is clear evidence of increased rates of weathering where levels of pollutants are higher, of damage to building materials and the uptake of pollutants by building materials. The building microclimate is, however, an important factor in determining the level of damage. There is a materials exposure programme (ICP Materials) run across Europe, Canada and the USA through the UNECE. Experiments under this programme using Portland stone showed that weight loss of samples exposed in London was more than 25 per cent greater than the loss of samples put into rural settings. On buildings themselves, it seems that surfaces which look clean may actually be experiencing more damage (loss of material) than those which look dirty where protective crusts build up. Some of the literature on air pollution on buildings has been summarized by Brimblecombe (2003).

4.1.5 Damage thresholds – critical loads and critical levels

While evidence of long-term increases in the acid loading of the atmosphere and of ecosystem change/damage accumulated from individual sites, it was realized that having clearly defined pollution thresholds for species or ecosystems would be a useful tool. The idea of thresholds was developed in Scandinavia and Canada in the 1970s and defined at the Stockholm environment conference in 1982. The term threshold was subsequently replaced by the terms critical load (for depositions) and critical level (for gas concentrations). A critical load (CL) has now been defined as 'a quantitative estimate of exposure to one or more pollutants below which significant harmful effects on specified sensitive elements of the environment do not occur according to present knowledge'.

The first workshop on critical loads was held under the auspices of the UNECE at Skokloster in 1988 (Nilsson and Grennfelt, 1988). Although the way in which critical loads are calculated has changed over time, the concept underpins the development of emissions reduction policies within the UNECE, the EU and nationally. The CL reflects the estimate of a receptor's (soil, water, vegetation type) buffering capacity, immobilization and uptake of deposited S and N. The lower the CL load, the more sensitive the site. Where deposition is greater than the CL, then the CL is said to be exceeded and damage is assumed to occur. The success of emissions reduction policies is gauged by the extent to which CL exceedance is reduced. CL can be defined for S only (acidity), N only (eutrophication) and S and N together (total acidity). Within the UNECE, CL are calculated at the national level and then submitted to the Coordinating Centre for Effects (RIVM, the Netherlands). Reports and details of methods are available through the CCE website (http://www.arch.rivm.nl/cce). If no national estimates are supplied, then CL are assigned by the Centre.

Early critical loads maps were described as empirical (also called level 0) and made an assessment of sensitivity based on existing data, most particularly the mineralogy of forest soils – these maps were essentially a reflection of underlying geology. The majority of CL estimates are now based on steady state or simple mass balance (SMB) models (level 1) which are equilibrium models assuming that systems respond very rapidly to changes in deposition. SMB models can be developed for individual ecosystems or habitat types (e.g. coniferous woodland) or for combinations of these. Details of the UK's current approach to calculating critical loads are available through the national critical loads mapping website (http://www.critloads.ceh.ac.uk). Reliable and consistent mapping of habitat distributions has become increasingly important in calculating CL. The UK now uses Broad Habitats derived from the UK Biodiversity Action Plan or ENNIS habitat classes. There is some CL work which uses dynamic models (level 2) i.e. those which allow for change through time, but the data requirements of such models are high and they are of limited applicability at large spatial scales. CL for freshwaters have developed in parallel with those for terrestrial ecosystems (CLAG, 1995).

Lake chemistry across Northern Europe was surveyed in 1995 and compared with the CL for S acidity. The highest percentage of lakes where deposition exceeded CL occurred in Norway (27 per cent), followed by the Russian part of the Kola peninsula (17 per cent). For Northern Europe as a whole it was estimated that CL exceedance was being experienced by about 22,000 lakes, 14 per cent of the total (Henriksen et al., 1998). CL derived by the different methods (empirical, steady state, dynamic) will each give a slightly different picture of ecosystem sensitivity. A sample CL for the maximum critical load for sulphur (5th percentile) for the EMEP area is illustrated in Fig. 4.6.

The science of CL is most highly developed in Europe, while the application of the approach in North America has been more restricted. The development of CL for southeast Asia has made more progress. The steady state mass balance approach (see above) has been used by the Coordinating Centre for Effects to calculate critical loads for acidity (S only) for China, Korea, Japan, the Philippines, Indo-China, and the Indian subcontinent. The resulting map can be seen at http://www.iiasa.ac.at/research/tap/rains_asia.docs. A global assessment of ecosystem sensitivity for terrestrial ecosystems has been made by Cinderby et al. (1998). Information on mapping and modelling acid deposition effects at the global scale are available through the Stockholm Environment Institute website (http://www.york.ac.uk/inst/sei/rapidc2/rapidc.html). A global analysis of critical loads exceedance of natural terrestrial ecosystems has been published by Bouwman et al. (2002).

Although CL have been widely adopted as a means of assessing ecosystem damage, they do have a number of problems and have been subject to criticism. Perhaps most important is the fact that, in most cases, it is difficult to identify a clear relationship between the exceedance of the CL and actual ecosystem damage. The clearest case is the empirical CL model for freshwaters where there is a response from a group of organisms (diatoms). In other instances, the ecological significance of CL exceedance is not well established. The move to steady state models has also meant that many key values have to be defined, or assigned.

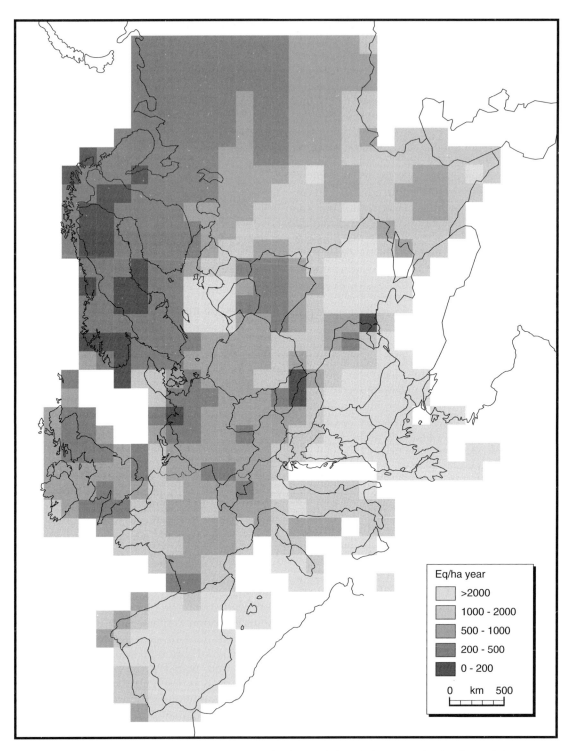

FIGURE 4.6 Sample CL for the maximum critical load for sulphur (5th percentile) for the EMEP area
(Source: Coordinating Centre for Effects, RIVM (ICP Modelling and Mapping))

Power generation companies have been particularly vocal in their criticism of the approach, both in terms of the underpinning science and the level to which that science has been open to public scrutiny (Skeffington, 1999). CL have, however, played a key role in the development of protocols within the framework in the LRTAP convention and are being used by the EU for directives related to acidification and eutrophication. In the development of these policies, CL data are combined with output from long-range transport models to predict changes in deposition and hence CL exceedance under a range of possible emissions scenarios. Some of these transport models are described in Chapter 6. The policy context for CL is discussed more fully in Chapter 7.

4.2 Heavy metals

In common with the precursors of acid deposition, heavy metals such as arsenic, lead and mercury, may come from a range of natural and anthropogenic sources. The major anthropogenic sources are fossil fuel combustion (especially coal), industrial processes (e.g. iron and steel industry) and waste incineration. Natural sources of metals include volcanoes, sea salt and soil dust. As a result of industrial processes, metals are released into the atmosphere as particles which may then be inhaled or ingested by a variety of means. Combustion may also yield metals as vapour. Three heavy metals have been identified as being particularly harmful: lead (Pb), cadmium (Cd) and mercury (Hg) (UNECE, 1998). Clear human health and other environmental impacts have been identified. Lead poisoning affects the production of haemaglobin and can result in irritability, stomach pains and, ultimately, brain damage. Although often bound to sediments, and hence immobilized, more soluble forms of these metals (such as monomethyl mercury) are extremely toxic and can bioaccumulate through the food chain. Acidification of soils and freshwaters can result in the release of metals formerly bound to the surface of particles. This occurs mainly when the pH reaches 4.5 or less.

Perhaps the most notorious heavy metal is mercury (Hg) which in its most toxic soluble form, methyl mercury (MeHg[1]), can cause damage to the central nervous system and finally paralysis. Apart from occupational exposure to mercury (working in mining or exposed to mercury vapour), most human exposure to mercury comes through the consumption of fish and other seafood where mercury can accumulate. The dangers of exposure were revealed in 1953 in Minimata in Japan where the use of mercury in a local factory led to contamination of the local fish stocks and severe health effects. Although the link to mercury from the factory was suggested in 1958, it was not officially acknowledged as the cause by the Japanese government until 1968. International limits for concentrations of Hg in fish are between 0.4 and 1.0 mg Hg per kg fish. Excess concentrations have been found in fish in south and central Finland, much of Canada (e.g. Ontario), Sweden and Maine and Wisconsin in the USA (Lindqvist et al., 1991). Poisoning can also occur where mercury is used as a fungicide for seed grains; 500 people died in Iraq in 1972 as a result of this. The areas around metal smelting industries suffer extreme local impacts. Notorious examples of this are the Kola Peninsula of north-west Russia and the area of Sudbury (near Ontario) in Canada, both centres for copper-nickel smelting. Both regions have also suffered severe damage from SO_2 emissions (see Section 4.1.5 above) as well as the effects of heavy metals. There has been a major clean up operation in Sudbury in recent years.

4.2.1 Emissions of heavy metals

Global estimates of emissions of heavy metals to the atmosphere suggest that in many cases anthropogenic sources are far more important than natural sources (see Table 4.3). In many urban areas more than 90 per cent of trace metals in the atmosphere come from human activities. Most emissions are from the Northern Hemisphere.

TABLE 4.3 Global emissions of trace metals to the atmosphere in 10^9 g yr^{-1} (Nriagu, 1991)

Element	Natural	Anthropogenic
Arsenic	12	19
Cadmium	1.4	7.6
Copper	28	35
Lead	12	332
Mercury	2.5	3.6
Vanadium	28	86
Zinc	45	132

Lead

The majority of the lead emitted into the atmosphere is in the form of unburnt, or partially combusted, fuel vapours from petrol vehicles. This organic lead is present as tetraethyl lead or trimethyl lead. There is a much smaller amount of inorganic lead particulates which result from burning some types of coal (with a high lead content) and from lead-processing industries. While organic lead passes through the body quite quickly, inorganic lead can accumulate in the body over long periods, replacing calcium in bones and teeth.

Cadmium

Non-ferrous metal industries are the largest human source of cadmium, followed by fossil fuel combustion (particularly the burning of brown coal) and waste incineration. Phosphate fertilizers can also be a source of cadmium. At the level of personal exposure, cigarette smoking is a major determinant of an individual's body burden of cadmium. Cadmium can be both inhaled and ingested and accumulates in the body. It is known to cause kidney damage.

Mercury

Estimates of natural emissions of mercury are highly variable as a result of the scarcity of measurements and the problem of re-emission. The figure given in Table 4.3 is low compared with a number of other published figures (up to 100 10^9 g yr^{-1}). Relatively little is known about the species of mercury emitted from different source types (Fitzgerald *et al.*, 1997). Close to sources, most pollution is probably associated with ionic and particulate forms of Hg, while in remote areas elemental Hg is probably more important. Mason *et al.* (1994) have made estimates of the pre-industrial and present global mercury cycle. They calculate that about 57 per cent of the total flux to the atmosphere now comes from direct anthropogenic emission (about 20 Mmol yr^{-1}) and 30 per cent from the oceans (10 Mmol yr^{-1}). The current atmospheric concentration is about three times the pre-industrial level. Over the last ten years anthropogenic emissions have declined in North America and Europe, but have increased in India and China (Seigneur *et al.*, 2001). There is a slow recycling of Hg from the oceans to the atmosphere and that held in soils is slowly released into lakes, rivers and coastal waters. As a result of these stores, there would be a time lag between any reduction in the emissions of mercury and reduced concentrations in the environment.

Although the focus here will be on trace metals in the atmosphere and their deposition, it should be borne in mind that discharges to water are generally much higher than those to air. In many cases, the build up of metals in the biosphere exceeds their natural removal rate by tenfold or more. As a result, in many ecosystems, inputs of trace metals from human activities are dominating the natural biogeochemical cycles. The long-range transport of heavy metals has been illustrated by studies of fluxes into the world's oceans. Higher fluxes occur close to the source regions, with the North Atlantic being the most polluted and the South Pacific the least polluted (WMO, 1989).

As with emissions of S and N compounds, signatories to the UNECE Convention on Long Range Transboundary Air Pollution must supply estimates of their emissions to air of heavy metals. Maps of emissions have been compiled for Europe (EMEP area) on a grid scale of 50 km × 50 km (Ryaboshapko *et al.*, 1999) and are also available through the EMEP/MSC-E website (http://www.msceast.org/hms). The highest emissions occur in industrial regions such as north-west Germany, southern Poland and the eastern Ukraine. Lead emissions show the most uniform distribution as they are largely associated with road transport. Cadmium emissions reflect their sources in non-ferrous metallurgy and fuel combustion and most come from the Russian Federation, Italy, Romania and Spain. Anthropogenic mercury is largely associated with coal consumption with the largest emissions coming from Poland and Germany. Mercury previously deposited can also be re-emitted from soils and surface waters. For the EMEP area it has been estimated that this source is about 10 per cent of the size of direct anthropogenic emissions. Since 1990, the emissions of heavy metals have declined: lead by 63 per cent, cadmium by 47 per cent and mercury by 52 per cent.

In the USA, lead is specified as a criteria pollutant with National Air Quality Standards. Emissions are now dominated by industrial processes (mainly metal processing) whereas in the past, motor vehicles were the major source (EPA, 2003a). In common with other developed economies, lead emissions have declined steeply since the 1970s; in the USA the reduction between 1975 and 1997 was 98 per cent. Since better data became available in 1988, the largest change has been in emissions from petrol engine vehicles as a result of the introduction of unleaded petrol.

In the UK, the NAEI reports emissions data for 13 heavy metals. Most of the emissions are estimated using UK specific emissions factors expressed as mass of metal emitted per unit mass of fuel burnt or unit mass of product. There is considerable uncertainty in these emissions estimates as the metal content of fuels and raw materials varies. The best estimates are probably for cadmium, lead and mercury (but still about ± 30 per cent) and the worst for beryllium, manganese and tin. Trends in UK emissions of cadmium, lead and mercury based on Dore *et al.* (2003) are summarized in Table 4.4.

Emissions of all reported metals have declined very significantly since 1970; for cadmium, lead and mercury this decline has been by 81 per cent, 97 per cent and 80 per cent respectively. The greatest change occurred in lead emissions between 1985 and 1986 as a result of changes in the lead content of leaded

TABLE 4.4 UK emissions of selected heavy metals (in tonnes yr^{-1})

	1970	1980	1990	1995	2000
Cadmium	26.8	20.6	20.3	12.0	7.2
Lead	7339	8151	2780	1535	193
Mercury	44.9	35.2	31.6	19	8.8

(Data from Dore *et al.*, 2003)

petrol. Leaded petrol was withdrawn from general sale at the end of 1999. The decline in coal burning, and controls on emissions from waste incinerators, have contributed to the overall reduction in emissions of these heavy metals. It is interesting to note, however, that based on new data supplied by the major power generators, estimates of emissions of mercury from coal fired power stations have been doubled (Goodwin *et al.*, 1999). Maps of heavy metal emissions are available and show quite different spatial distributions for pollutants such as mercury, where point sources are significant, compared with lead where point sources are not apparent on the maps, but major roads are.

The UK has a heavy metal monitoring network with 15 sites measuring the content in rain, particulates and clouds. Information on the network is available from http://www.heavymetals.ceh.ac.uk/info.htm.

4.2.2 Chemical transformation of heavy metals

In common with other pollutants, the physical and chemical forms in which heavy metals occur determines their transport, transformation and removal from the atmosphere. The form in which mercury is emitted to the atmosphere varies with source type. The majority is in the gas phase, with a smaller proportion emitted directly as particles. Most is emitted as elemental Hg (written as Hg^0), but waste incinerators, for example, emit Hg mainly in the form of divalent inorganic compounds which are easily dry deposited. The main chemical transformation of mercury is aqueous phase oxidation (Petersen *et al.*, 1995). The resulting dissolved oxidized mercury in water droplets is adsorbed on to soot particles, so a significant proportion of Hg in precipitation is in the particulate phase. Any particulate Hg is effectively dry deposited or wet deposited following scavenging within or below clouds. Ozone is the major oxidant of Hg, so the concentration of secondary particulate Hg is related to the concentration of O_3 and carbon particles as well as Hg. Once deposited, reactive Hg leads to the production of soluble, toxic monomethyl Hg in natural waters.

Natural sources of lead and cadmium are wind erosion and volcanic activity and they are present on large particles. Lead and cadmium from anthropogenic sources (such as vehicle exhausts) are present on sub-micron size aerosols (small particles). It is believed that these metallic particles do not undergo transformation in the atmosphere, but can be dry deposited or washed out (wet deposited).

4.2.3 Long-term trends

4.2.3.1 Ice cores

As with acidifying pollutants, long-term records of changing heavy metals concentrations come from ice and lake sediment cores. The ice core record of heavy metal concentrations has been reviewed by Boutron *et al.* (1994). The longest records come from the Antarctic and extend back beyond the last interglacial period (> 150,000 years). The natural variability in heavy metal concentrations in these cores is a function of climate, with high levels of lead, cadmium and zinc occurring around the last glacial maximum as a result of enhanced fluxes of soil and rock dust in dry, windy conditions. Studies of a 40,000 year record of mercury (Hg) from Dome C in the Antarctic show a similar pattern to the other metals (much lower concentrations in the Holocene (the last 10,000 years) than those recorded around the last glacial maximum. In this case, however, elevated Hg levels during the glacial are probably associated with the release of more mercury from the more productive glacial oceans (Vandal *et al.*, 1995).

Ng and Patterson (1981) studied the lead content of ice cores from Camp Century in Greenland and Byrd Station in Antarctica. As well as illustrating very serious contamination following sampling of the outer surfaces of the ice cores, their study showed very low levels of lead (< 1 ng/kg) in ice more than 3000 years old in Greenland and more than about 2000 years old in the Antarctic. Hong *et al.*'s (1994) study of Greenland ice showed evidence of early lead pollution resulting from Greek and Roman smelting activities. Greenland ice from the 1960s showed a 300-fold increase in lead concentration relative to this background, with the most rapid increase occurring over the last 250 years. The ratios of lead isotopes have also been used to try to fingerprint the sources of this early pollution and southern Spain has been identified as one of these sources (Rosman *et al.*, 1997). The Antarctic ice showed a tenfold increase in Pb concentration over the last 100 years. Evidence for enhanced levels of Cd from human activities in Greenland ice, however, seem to be confined to the period since the nineteenth century (Hong *et al.*, 1997).

More recent work on a core from Summit (Greenland) studied the concentrations of a range of heavy metals in snow and ice accumulated between 1967 and 1989. This showed clear evidence for a decline in inputs of Pb, Cd and zinc (Zn) from anthropogenic sources, in line with decreasing emissions of these pollutants. The decline was most dramatic for lead, which declined 7.5-fold over the period of record – this change was attributed to the move away from leaded petrol. The patterns of change in Pb in the Greenland and Antarctic cores are summarized in Figure 4.7.

Using the ratio of two isotopes of lead, ^{206}Pb and ^{207}Pb, it is possible to fingerprint the sources of the anthropogenic lead reaching Greenland. The main source areas with distinctive 'fingerprints' are the USA, Canada and Europe. Through the 1970s, the USA was the dominant source of Pb reaching Greenland (> 60 per cent). This dropped dramatically to only 25 per cent by 1988, while Europe and Canada became more important sources. This largely reflects the earlier implementation of measures to control the lead content of fuels in the USA

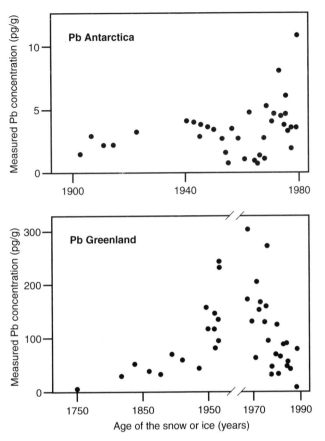

FIGURE 4.7 Changes in Pb concentrations in snow in the Antarctic (since 1900) and Greenland (since 1750) (Based on Boutron *et al.* (1994), Fig. 5)

(see Chapter 7). The only metal to show no significant trend in the Greenland cores was copper (Cu), a large proportion of which is believed to come from natural sources.

There is no clear evidence of enhanced mercury (Hg) deposition to Antarctica over the last 200 years. The oceans still seem to be the major source of Hg deposited here. A general review of the significance of contamination of remote areas due to long-range atmospheric transport and deposition of mercury produced by human activities has been carried out by Fitzgerald *et al.* (1997). The oceans, local geology and diagenetic processes all produce locally variable Hg levels but cannot adequately explain the observed spatial and temporal changes in mercury concentrations in ice cores, lake sediments, peats and soils. Atmospheric inputs of

mercury resulting from human activities must be responsible and it is known that the land is the major sink for atmospheric Hg. It has been estimated that over the last 100 years anthropogenic emissions have tripled the concentrations of Hg in the atmosphere and oceans.

4.2.3.2 Lake sediments

Changes in atmospheric concentrations of heavy metals are clearly recorded in lake sediments, although the individual catchments may also reflect local mining and smelting activities. Renberg *et al.* (1994) reported a study of the Pb content of cores from 19 lakes across Sweden. Their study showed that Pb levels went above background about 2600 years ago, with a small peak about 2000 years ago. A more significant increase began about 1000 BC and accelerated through the nineteenth and twentieth centuries to a peak about AD 1970. The authors suggest that the earliest lead peak was associated with extensive lead production during the Roman period, and this accords with the evidence for long-range transport of lead pollution provided by the ice cores.

In another study, Renberg (1986) reports on laminated lake sediments from seven sites in northern Sweden. Two of the lakes recorded the effects of local smelting, but the others showed changes in Pb, Cd and Hg since the nineteenth century. The concentration of lead in the soluble sediment fraction was shown to have more than doubled from the early nineteenth century through to the mid-twentieth century, followed by a recent decline. Cadmium concentrations showed a continuing increase. Increases in mercury did not begin until the twentieth century, mainly in the latter half – the concentration of mercury in the soluble sediment fraction of recent sediments was 4 to 10 times that of sediments from the eighteenth and nineteenth century. The association of the heavy metals with combustion sources was shown by the corresponding increases in the numbers of carbonaceous particles. A study of lake sediment cores from Loch Lomond, in western Scotland, showed that lead levels had increased sharply from the late eighteenth century. Until the early 1930s, the ratio of $^{206}Pb/^{207}Pb$ was typical of emissions from heavy industry and coal combustion (i.e. enriched in ^{206}Pb). The ratio then changes as emissions from vehicle exhausts (depleted in ^{206}Pb) became more important. In common with other records, Pb fluxes in the second half of the twentieth century were lower (by some 40 per cent) than those of the first half (Farmer, 1994).

Studies of lake sediment cores have also contributed to considerations of the overall atmospheric loading of mercury. Many studies have shown local decreases in mercury inputs to lakes as a result of the cessation of local mining and smelting activities. A study by Engstrom and Swain (1997) compared records from south-eastern Alaska (effectively a background site, remote from anthropogenic sources) with sites from Minnesota in the mid-west of the USA. The Alaskan records indicated continued rising deposition of Hg through to 1993, while the mid-western sites showed a decline which was apparently a response to reduced regional mercury emission and deposition.

4.3 Tropospheric ozone

Ozone was initially discovered by Schönbein in the mid-nineteenth century. Unlike the precursors of acid deposition or heavy metals, ozone (O_3) is not emitted into the atmosphere, but is a product of photolytic and photochemical reactions. The vast majority of ozone (about 90 per cent) occurs in the stratosphere (see Chapter 3) where it is produced as a result of photolysis of molecular oxygen by UV radiation. Here the focus is tropospheric ozone.

For many years the ozone in the troposphere was thought to be chemically inert. It was assumed to have originated in the stratosphere and to be lost by deposition to the surface. It is now clear that this view is far from correct (Crutzen et al., 1999). Tropospheric ozone is largely produced by photochemical reactions involving oxides of nitrogen (NO_x) and volatile organic compounds (VOCs). Carbon monoxide (CO) and methane (CH_4) also play an important role. As described above, ozone is an important oxidant, but in an unpolluted atmosphere does not accumulate. In areas downwind of high concentrations of its major precursor pollutants, NO_x and VOCs, and in the presence of sunlight, very high concentrations of O_3 can be generated. These areas are usually around urban and industrial conurbations where ozone forms an important element of photochemical smog. This phenomenon was first noted in the Los Angeles Basin in the 1940s (see Chapter 5) and the mechanism for its formation identified by Arie Haagen-Smit in the early 1950s. As described in Chapter 3, plumes of high concentration can travel hundreds of km from the source region. Monitoring off the eastern seaboard of the USA has shown concentrations of up to 150 ppb below 2 km in the summer

TABLE 4.5 Selected UNECE VOC emissions in 1000s tonnes yr^{-1} (Data from Vestreng, 2003 and US EPA Technology Transfer Network)

	1980	1985	1990	1995	2000
Belgium	274*	274*	274	262	233
Czech Republic	275*	275	441	292	227
Germany	3224	3190	3220	2021	1605
Italy	2179	1992	2041	1800	1557
Norway	173	231	294	367	367
Poland	1036	1011	831	769	599
Canada	2099	2851	2997	2639	2493
USA	23596	21978	19037	18676	17875

*expert estimate

months (WMO, 1999). High concentrations of tropospheric ozone, however, have been associated with a range of environmental impacts including damage to vegetation and building materials, and health effects. Ozone is also a greenhouse gas (see Chapter 3). As a secondary pollutant, policies designed to reduce both ambient and peak ozone concentrations have to be more sophisticated than those targeting primary pollutants. The timescales for the chemical production of ozone and the transport distances of the pollutants involved make tropospheric ozone a regional and indeed hemispheric problem.

4.3.1 Precursor emissions

As described above, O_3 is a secondary pollutant, but trends in the amounts and distributions of the main precursor primary pollutants, NO_x and NM-VOCs, can be traced. (The quantities, trends and sources of NO_x have been outlined above.) Emissions of NM-VOCs across the UNECE area are tabulated in non-speciated form, and some of the trends (for the same countries shown in Table 4.1) are shown in Table 4.5.

The trend in emissions has generally been down, but as the figures for Norway indicate emissions from some countries are increasing.

In some cases biogenic emissions of VOCs can be significant and exceed anthropogenic emissions. In the USA, isoprene (mainly from oak forests) and monoterpenes (from pine, citrus and maple, for example) are regionally dominant. Haze in the Smoky Mountains National Park has been attributed to monoterpenes from the natural vegetation. Total biogenic emissions from vegetation in the USA are probably higher than those from anthropogenic sources. Anthropogenic emissions come mainly from industrial sources (51 per cent), followed by transport (40 per cent). These emissions declined by 20 per cent between 1988 and 1997.

Data for the UK are available in NAEI reports (e.g. Dore et al., 2003) and illustrated in Table 4.6. Emissions of NM-VOCs showed an overall increase from 1970 to 1989, but there has been a steady decline over the 1990s.

The spatial distribution of emissions is strongly influenced by population, as solvent use and vehicle emissions are higher in urban areas. The largest sources of NM-VOCs are

TABLE 4.6 UK NM-VOC emissions in 1000s tonnes yr^{-1} (Dore et al., 2003)

1970	1980	1990	1995	2000
2172	2338	2603	2149	1596

road transport (19 per cent) and solvents (29 per cent). Road transport emissions peaked in 1989 and have decreased since then due to the introduction of catalytic convertors and the switch to diesel. Looking in more detail, the individual VOCs making the greatest contribution to the total are butane, ethanol, ethane and propane.

4.3.2 Production, destruction and deposition

The reactions involving NO, O_3, NO_2 and sunlight in the photostationary state have been described in Section 4.1.2 above. Net photochemical production of O_3 occurs when peroxy radicals (HO_2 and RO_2) are generated which can then *et al* NO to NO_2 without the consumption of O_3 effectively disturbing the photostationary state:

$$NO + HO_2 \rightarrow NO_2 + OH$$

These peroxy radicals can be produced by the oxidation of CO, CH_4 or non-methane VOCs (NM-VOCs) by OH. In these reactions, OH is regenerated. Although globally the majority of OH reacts with CO and CH_4, at the regional/local scale reactions with NM-VOCs from anthropogenic sources (such as solvent use and road transport) become dominant. The peroxy (free) radicals produced by the oxidation of these hydrocarbons enable O_3 concentrations to build up in urban areas where they would otherwise be low. The number of oxidation cycles the OH/peroxy radicals can go through before reaching terminating reactions (i.e. final sinks for the radicals) dictates how much excess O_3 can be produced. The number of cycles is described as the 'chain length' for ozone formation (PORG, 1997), and efforts to reduce ozone concentrations must shorten these 'chain lengths'.

It is possible to predict O_3 concentrations for different initial concentrations of NO_x and VOCs (and sunlight). Where the ratio of NO_x to VOC is high, the system is said to be VOC-limited, while for low NO_x to VOC ratios, the system is NO_x-limited. In a VOC-limited system 'chain length' is shortened by reducing VOC emissions. However, if policy brought about a reduction in NO_x, but not VOCs, in a VOC-limited area, then O_3 concentrations would increase (longer 'chain length'). The implementation of policies for ozone reduction is, therefore, a complex process, especially as the definition of a region as VOC- or NO_x-limited may vary depending on wind direction in relation to pollution sources. The implications of demands to reduce NO_x emissions in order to address problems associated with acid deposition should also be borne in mind.

The largest sink for tropospheric ozone is dry deposition to terrestrial surfaces. A small proportion will be lost through uptake into cloud water droplets.

4.3.3 Current patterns

Background ozone concentrations have been measured across the EMEP area since 1984, with data being collected systematically since 1988. Initially, coverage of southern Europe was extremely poor, although this area might be expected to have high summer O_3 concentrations. This has now improved and in 2003 there were about 1800 monitoring sites reporting to the EEA (Fiala *et al.*, 2003). Annual average O_3 concentrations are about 20–25 ppb (40–50 µg m^{-3}) on the northern and western fringes of Europe and 30–35 ppb in central Europe (Dollard *et al.*, 1995). Almost all sites, however, experience concentrations in excess of 40 ppb (80 µg m^{-3}) at some time and all sites with an annual average of > 30 ppb have a summer monthly mean above 40 ppb. Summertime ozone episodes, with concentrations of 100–150 ppb (200–300 µg m^{-3}), are a regular occurrence across Europe and are discussed further in Chapter 5. Although the trend in peak O_3 across Europe is down, there has been little change since 1995 in reported exceedances of ozone threshold values (see Section 4.3.5). Information on ground level ozone concentrations in the EU is available from the website at http://www.air-climate.eionet.eu.int/databases/o3excess.

Ground level O_3 has been described by the EPA (1998) as a 'pervasive pollution problem throughout the United States'. The highest ozone concentrations usually occur in southern

California, the Gulf Coast, the north-east and north-central parts of the USA. Concentrations are highest in suburban areas, followed by urban, then rural. The highest 1-hour concentrations (based on the second daily maximum 1-hr concentration) have historically been recorded in Los Angeles, although in 1997 (when the summer in California was unusually wet) this 'distinction' went to Houston, Texas.

Since 1988, ozone concentrations have been measured at a network of 660 monitoring sites. Over the period 1988–97, 1-hr and 8-hr ozone values fell by about 19 per cent, with the biggest decline in the highest concentrations. The 20-year trend (based on fewer monitoring sites) shows a decline in maximum 1-hr values (second highest as above) of about 30 per cent since 1978 from about 150 ppb to just over 100 ppb. Most non-urban sites, however, show no significant trends from 1980 to 1995. Peak ozone values are highly dependent on meteorological conditions and the US data show clear high ozone years in 1980, 1983, 1988, 1995 and 2002. The EPA have produced a model for 41 urban areas which attempts to account for the variability induced by meteorology and this suggests a decline in O_3 of about 1 per cent per year since 1986 in these urban areas. Regionally, the biggest improvements in O_3 since 1983 have occurred in the north east and Pacific south west regions of the EPA where decreases in the 1-hr maximum have exceeded 20 per cent. In spite of these improvements, the EPA's most recent 8-hr O_3 primary standard for health and welfare (see Chapter 7) is, on average, exceeded. In 1997, 48 million people lived in counties where the 1-hr maximum exceeded the 1-hr NAAQS (EPA, 1998). Eight-hour ozone values have also generally declined since 1983 except in the Pacific north west (EPA, 2003b).

There is a lot of concern in the USA about air pollution, particularly degraded visibility caused by photochemical smog, in remote, wilderness areas. As a result, O_3 monitoring has been carried out in National Parks (defined as Class 1 areas under the Clean Air Amendments 1977 – see Chapter 7). The data show no significant overall trends from 1988 – 1997, but two individual sites, both in the Great Smoky Mountains National Park, had statistically sig-

nificant increases in O_3 levels. The importance of natural VOCs in this area should, however, be borne in mind. Aerosols, both primary and secondary, are important contributors to poor visibility. In the USA, secondary sulphate aerosols predominate in the east, while primary particulates are more important in the west.

Ozone concentrations across the UK are measured through both a rural and urban monitoring network. Measurements in the UK started in 1972. The highest mean concentrations occur in rural areas and, on an annual basis, are 5 to 10 ppb lower in urban areas. The annual average concentrations in rural areas lie between 20 and 30 ppb (40–60 $\mu g\ m^{-3}$), while hourly values can range from 0 to 150 ppb (PORG, 1997). Ozone shows clear diurnal and seasonal trends. In the UK, highest background O_3 levels are recorded in the spring. The situation in urban areas is described in Chapter 5. National maps of summer ozone concentration are corrected for both topographic and urban effects. The resulting maps show low levels around conurbations, with the highest values in southern and western England. Annual concentrations at the rural sites have shown either no change or a small increase over the period 1986–95. Maximum hourly values (during summer photochemical episodes) have, however, shown a clear decline. Hourly average concentrations in the period 1972–85 often exceeded 200 ppb, while more recently values have rarely exceeded 150 ppb. Even in the significant ozone episode of summer 2003, the highest concentration was 125 ppb (250 $\mu g\ m^{-3}$).

4.3.4 Long-term trends

Direct, high quality measurements of ozone have only been made over the last 30 years or so. There are, however, some measurements dating back to the late nineteenth century which help to piece together the long-term trends in O_3 concentration. The two main historical data sets come from Montsouris near Paris (1876–1907) and Moncalieri near Turin (1868–1893) (Anfossi et al., 1991). These early data have to be manipulated to give ppb values which can be compared with modern measurements. The annual mean concentration in northern Italy was about 8 ppb last century, while modern values for semi-rural and remote sites in the same area are about 25 ppb. This increase in concentration has occurred not just at the surface, but throughout the troposphere. Seasonal variability has also increased greatly. A more recent long data set comes from Hohenpeissenberg, Germany where measurements started in 1967. Across Europe as a whole, ground level mean O_3 concentration seems to have increased from about 10 ppb to 30 ppb. This reflects increasing emissions of the precursor pollutants: NO_x, VOCs, CO and CH_4.

4.3.5 Damage due to ozone

There is good evidence for ozone damage to vegetation and a range of different types of materials (e.g rubber, paints), as well as health effects. These findings have been based on a range of approaches including field and laboratory experiments and epidemiological studies. The damage is caused because ozone is an oxidant, reacting with organic molecules. Ozone damage to plants occurs when the gas is able to enter through the stomata, which is usually during the day. Due to the moisture in the plants, hydrated forms of ozone are rapidly formed and these damage cell membranes or plasma. Damage to the cell membranes effectively changes their permeability and potassium ions, sugars and amino acids can be lost. Visible injury to foliage results with the development of pale flecks or stipples, and the overall growth and viability of the plants can be affected.

A great deal of attention has been focused on the response of agricultural crops to ozone exposure. Between 1987 and 1991, nine European countries participated in a programme to expose a variety of crops to different ozone concentrations in open topped chambers (PORG, 1997). These experiments showed that yield decline could be correlated with cumulative exposure over a threshold of 40 ppb (known as AOT_{40}). Using data from a range of experiments, the relationship, between AOT_{40} and yield loss for spring wheat was used to set

a critical level for agricultural crops, based on exposure over 40 ppb during the three months when the crops are most sensitive to ozone (May – July in northern Europe). The critical level for forest trees is based on data for beech (possibly the most sensitive). AOT_{40} is again the basis for calculation, but with values calculated over daylight hours of a six-month growing season. Maps of ppb hours above AOT_{40} for the UK, show exceedance of the damage threshold for both crops (especially central and southern England) and forest (southern Britain).

The health effects of ozone are relatively well understood, although the periods of highest ozone concentrations (in the summer) are also associated with elevated levels of other photochemical pollutants such as acid aerosols. Ozone can penetrate deep into the lungs and will cause inflammation. It may be that exposure to ozone also increases an individual's sensitivity to other allergens.

In 1992, the Directive 92/72/EEC on air pollution by ozone was adopted by the EU. The directive identified a set of thresholds related to different receptors and including thresholds for population information (180 μg m^{-3}) and warning (360 μg m^{-3}, 240 μg m^{-3} over three consecutive hours from September 2003) (see also Chapter 7). These thresholds are set out in Table 4.7. Across Europe, the warning threshold is only exceeded occasionally, but the information threshold is exceeded in all countries except Ireland, Denmark, Sweden and Finland.

The summer of 2003 (especially August) was extremely warm and long lasting and widespread high ozone concentrations were recorded across Europe. The number of days when the information threshold value of 180 μg m^{-3} (90 ppb) was exceeded are shown in Fig. 4.8.

In the UK, the Expert Panel on Air Quality Standards (EPAQS) limit for health effects from O_3 is an 8-hour moving average concentration of 50 ppb (100 μg m^{-3}). This value is exceeded quite regularly. The UK's Committee on the Medical Effects of Air Pollutants (COMEAP, 1998) reported that for Great Britain, assuming a threshold value of 50 ppb ozone (summer only), 700 deaths were brought forward and there were 500 additional or early hospital admissions for respiratory diseases. The UK

REGIONAL SCALE AIR POLLUTION 81

TABLE 4.7 Threshold ozone concentrations for different receptors within the EU

Threshold type	Concentration in $\mu g\ m^{-3}$	Averaging period (hrs)
Health	110	8
Vegetation	200	1
	65	24
Information	180	1
Warning	360	1

was also affected by high ozone levels in the summer of 2003. The maximum daily 8-hour mean over two weeks in early August was 104 $\mu g\ m^{-3}$ (52 ppb).

As discussed further in Chapters 5 and 7, O_3 concentrations across both Europe and North America currently regularly exceed the various damage thresholds set by different organizations. Measurements at remote sites also suggest that the Northern Hemisphere background concentration is itself increasing, further exacerbating the effects.

4.4 Regional, global or local?

The range of pollutants discussed here – oxides of sulphur and nitrogen, heavy metals and ozone – all have impacts at a range of scales. Sulphur and nitrogen oxides play a role in determining radiative forcing at the global scale (Chapter 3) and acidification (discussed here), as well as being pollutants close to sources and in urban areas (see Chapter 5). Nitrogen oxides play a role in determining the formation and destruction of ozone at different levels in the atmosphere. Ozone itself is an important component of the atmosphere at all scales (see also Chapters 3 and 5). Heavy metals have also been shown to have effects from the local to the regional. The distribution of the pollutants described here as 'regional' is changing. Emissions from the traditional source regions of North America and Europe have been declining over recent years, while those from Asia have (until recently) increased. The effects of the emissions changes are still being assessed and will be considered further through the use of models in Chapter 6. The policy context for these changes is described further in Chapter 7.

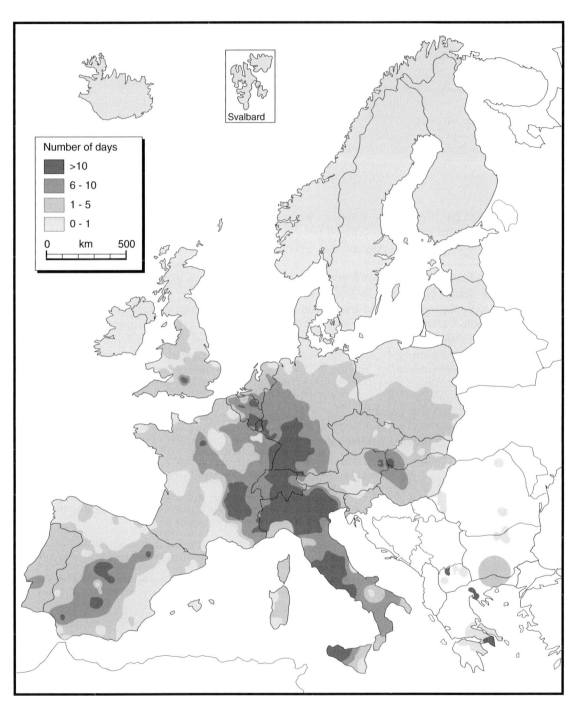

FIGURE 4.8 Exceedances of the ozone threshold value for informing the public (1hr > 180 μg m^{-3}) in the summer of 2003 (Source: Fiala *et al.* (2003), European Environment Agency)

5

Urban air quality

5.0 Introduction

In 1999, the world's population officially reached 6 billion (doubling since 1960), of whom about half live in urban areas. The rate of increase in overall population is fastest in the developing countries (the so-called 'south'), as is the rate of urbanization. In 1990, it was estimated that 70 per cent of the developed 'north' was urbanized, compared with 34 per cent of the 'south'; by 2025 it is believed that these figures will be 80 per cent and 57 per cent respectively. Urban areas, with their high population densities and high levels of energy consumption, are associated with a range of environmental problems of which air pollution is only one. Health effects are the primary area of concern related to urban air quality, resulting from exposure to pollutants both indoors and out. Patterns of exposure will be dictated by a variety of socio-economic factors as well as by the basic geography and meteorology of the cities themselves. Unfortunately, many major cities occupy sites where natural conditions restrict the dispersal of pollutants and hence increase concentrations.

5.1 Global urbanization

The future will see a marked shift in the geographical distribution of population and poten-tial pollution. Projections suggest that by 2050, 60 per cent of the world's population will live in Asia (with only 7 per cent in Europe). Although the global rate of population growth has slowed to 1.3 per cent (http://www.unfpa.org), about 90 per cent of this growth is in developing countries. The environmental stresses of rapid urbanization are exacerbated by higher population densities in the cities of the developing world. London has about 4,000 people per km^2, Beijing more than 27,000 per km^2, Mexico City about 34,000 per km^2 and Calcutta some 88,000 per km^2 (Elsom, 1996).

Population growth is matched by increasing energy consumption and vehicle ownership. The industrial nations of the 'north' currently use 40 per cent more energy in total than the developing nations (which include most of Asia, Africa, the Middle East, Central and South America). This pattern will change, with energy consumption in the so-called developing countries estimated to exceed that of the industrialized nations by 2020 (http://www.eia.doe.gov/emeu/international/contents.html). Global energy consumption is predicted to rise by 58 per cent between 2001 and 2025, with the largest growth in Asia (EIA, 2003). The number of vehicles has increased by a factor of 10 since 1950 and it seems likely that the rate of increase in vehicle numbers may exceed the rate of population growth in the future. In 1996, there were an estimated 676 million vehicles worldwide; this may rise to 1 billion by 2025. At present, the countries of the 'south' own only 10 per cent of vehicles, but the rate of rise is very rapid. In the 'north' the increase in vehicles is about 2–3 per cent per year; in the 1980s the equivalent rates were 11 per cent per year in Brazil, 14 per cent per year in China and 30 per cent per year in Korea. In 1990, China had a population of 1.2 billion, but only some 5 million vehicles and 700,000 private cars. This total number of vehicles was less than that in Los Angeles alone! The number of vehicles in China, however, is nearly doubling every five years, although levels of ownership (vehicles /1000 people) remain low (http://www.earthtrends.wri.org) click on Energy and Resources.

The 1992 UN Conference on Environment and Development (the Rio Summit) recognized that there was a need for immediate action on the environmental conditions in cities and the Agenda 21 action plan included specific recommendations relating to urban air quality. These recommendations included: developing adequate monitoring systems; providing access to clean technologies; access to training; better data collection and assessment and the development of management skills (UNEP/WHO, 1996).

UNEP/WHO (1996) have suggested that about 50 per cent of the world's urban population are exposed to harmful concentrations of SO_2 and particulate matter, with other adverse effects coming from NO_2, CO, O_3 and Pb. Any assessment of urban air quality requires good quality, reliable monitoring data (see Chapter 2). Since 1973, the WHO have been assisting in the collection of air quality data for urban areas, compiling the results and making them available, and providing technical training and advice. Between 1975 and 1996 this work was carried out with UNEP through the Global Environmental Monitoring System (GEMS/AIR). This programme has now been taken over by the Air Management Information System (AMIS), also developed by the WHO. Based on the GEMS/AIR data, UNEP/WHO published two reviews of global urban air quality. The first of these focused on the 'megacities' of the world (with populations in excess of 10 million) (UNEP/WHO, 1992) and the second on a group of cities with populations between 3 and 10 million (UNEP/WHO, 1996). Data collected through

AMIS are validated and quality assured, but cities and regions collect air quality data using their own, individual protocols. Other data reported using agreed criteria are collected by the OECD every two years. These data include results for two types of urban sites: traffic/commercial and residential. Air quality data for cities in western Europe, the USA and Japan for 1988–93 have also been published (OECD, 1999). Up-to-date air quality information for many urban areas is now available through the www from a range of government agencies.

5.2 Health effects of air pollution

The relationship between urban air quality and health came to the fore in the early and mid-twentieth century when a series of severe pollution episodes (see Section 5.5) resulted in demonstrable negative health effects. Episodes in the Meuse Valley in Belgium (1930), Donora in the USA (1948) and London in the UK (1952) were all clearly related to adverse health effects ranging from headaches and vomiting, to additional deaths. Detailed analyses of data from these episodes highlighted that the very young and the elderly, especially those with pre-existing cardiorespiratory disorders, were at most risk. Relationships between air pollution and health have been established through two basic types of studies: controlled exposures of volunteers and epidemiological studies which can relate outcomes (such as hospital admissions) to the concentrations of pollutants while taking into account other confounding factors such as the weather, socio-economic status and lifestyle. Health-based guidelines for a range of air pollutants have used data from both controlled experiments and epidemiological studies. The major effects of a range of air pollutants are summarized in Table 5.1 (WHO, 2000).

The WHO World Health Report 2002 suggested that particulate air pollution might account for 2 per cent of cardiorespiratory mortality globally, with developing countries bearing the greatest burden.

Health effects have been identified for a wide range of other pollutants including known

TABLE 5.1 Health effects of major air pollutants (WHO, 2000)

Pollutant	Effect
SO_2 (often considered with SPM)	Impaired function of airway and lungs.
SPM (including PM_{10} and $PM_{2.5}$)	Effects depend on particle size and concentration. Results in respiratory problems. Finer particles carry further into the respiratory system. May be no safe level.
NO_2	Increases bronchial reactivity, reversible and irreversible lung damage, may also affect the spleen and liver.
CO	Reduces oxygen carrying capacity of the blood as CO binds very efficiently with haemoglobin. Affects brain, heart and muscle. Developing foetuses very vulnerable.
O_3	Inflammation of airway, reduced lung function.
Pb	Affects haemoglobin production, central nervous system and brain function.

carcinogens such as arsenic, benzene, benzo[a]pyrene, chromium, nickel and PAHs (polycyclic aromatic hydrocarbons).

Developing clear relationships between concentrations of pollutants and health outcomes is complex, particularly in relation to long-term exposure (Beverland, 1998). Detailed studies which try to take into account as many factors as possible yield the most reliable results. The Six Cities study in the USA (Dockery *et al.*, 1993) followed more than 8000 adults in six cities for a period of 14 to 16 years. The study showed that there was a relationship between mortality and pollution, especially the level of particulates, when factors such as age, sex, occupation and smoking were taken into account. In the UK, the longest time series study (14.5 years) is of data for Edinburgh which showed that there was a relationship between black smoke concentrations and respiratory mortality for people aged 65 or over (Prescott *et al.*, 1998). As well as the problems of accounting for a wide range of potential confounding factors, two other important issues relate to inadequate exposure assessment and the possible synergistic effects of a mix of air pollutants. Concentrations of air pollutants are measured at a limited range of sites which will rarely provide an adequate record of any individual's actual exposure (at work, at home, travelling

etc). It is becoming more feasible to take personal exposure measurements but there are still considerable practical difficulties for long-term studies. The issue of whether combinations of pollutants have more health effects than single pollutants remains unclear and requires further investigation.

As health outcomes have become an increasingly important focus for air quality legislation (see Chapter 7), so the role of medical advice has grown. In the UK, the Department of Health take advice from the Advisory Group on the Medical Aspects of Air Pollution Episodes (AGMAAPE) and from the Committee on the Medical Effects of Air Pollutants (COMEAP) (http://www.advisorybodiesdoh.gov.uk/comeap). The findings of these bodies are then considered by the group set up to recommend air quality standards for the UK, EPAQS (Expert Panel on Air Quality Standards).

A report by COMEAP (Department of Health, 1998) examined current understanding of the relationships between concentrations of air pollutants and health effects (daily deaths and admissions to hospitals for respiratory diseases). The group was able to specify dose-response relationships between concentrations and health outcomes for particulates, ozone and sulphur dioxide; for NO_2 a coefficient was defined relating to hospital admissions. The

National Air Quality Strategy (Department of the Environment, 1997) and its subsequent amendments (2000, 2002 (England), 2003) reflect the policy application of our improved understanding of air pollution and its health and non-health effects, taking into account the costs of abatement and its benefits. Placing monetary value on health benefits has proved difficult (an assessment of willingness to pay to reduce deaths yielded estimates in the range of £2,600 to £1,400,000). It is clear, however, that current and proposed air quality control measures will result in substantial reductions in deaths brought forward and in hospital admissions (18,500 and 22,000 respectively between 1996 and 2005).

5.2.1 Setting global standards

Many countries (or country groupings such as the EU) will set their own air quality standards (see Chapter 7). In some developing countries, however, there may not be the money or expertise to establish such standards so globally accepted criteria will be helpful. Global standards also facilitate comparison between different countries.

The WHO first published air quality guidelines in 1964 relating to effects on humans, animals and vegetation. Health-based standards for a range of pollutants (SO_2, suspended particulate matter (SPM), CO and photochemical oxidants) were published in 1972. Then in 1987, the WHO produced their Air Quality Guidelines for Europe (WHO, 1987) for a wider range of pollutants. These guidelines were widely adopted outside Europe and effectively became the global standards (see UNEP/WHO, 1992). However, as the original guidelines were based only on epidemiological and toxicological results from Europe and North America, it became apparent that there was a need for a broader consideration of data from and conditions in developing countries. Revised WHO guidelines were developed in 1997 and published in 1999. These guidelines aim to protect human health by helping countries to set their own air quality standards while taking into account technological and cultural constraints. Guideline values, tolerance concentrations or risk assessments were published for a wide range of pollutants (WHO, 2000). The guideline values for some of the traditional air pollutants (SO_2, NO_2, CO, O_3 and Pb) are set out in Table 5.2. These values are based on different health outcomes. No guidelines were set for suspended particulate matter (SPM). It was recognized that most health effects are associated with fine particles (PM_{10} or finer), especially with ultra-fine $PM_{2.5}$. The WHO did not feel able to set a threshold for 'no effects', so no guidelines were stipulated. The 1987 WHO guideline for total suspended particulate (TSP) (determined gravimetrically usually by using a high volume air sampler) was 120 $\mu g\ m^{-3}$ over 24 hours.

TABLE 5.2 Selected WHO 1999 guidelines for 'classical' air pollutants

Pollutant	Guideline value $\mu g\ m^{-3}$	Averaging time
SO_2	50	1 yr
	125	24 hrs
	500	10 mins
NO_2	40	1 yr
	200	1 hr
CO	10000	8 hrs
	100000	15 mins
O_3	120	8 hrs
Pb	0.5	1 yr

5.3 The changing face of urban air pollution

With few exceptions, urbanization has been accompanied by industrial development, often with housing and industry being mixed together. As widespread industrialization relied on coal, urbanization was accompanied by smoking chimneys (industrial and domestic). Smoke and SO_2 were, therefore, the first major urban pollutants. As measures have been taken to eliminate these pollutants, at least in areas such as western Europe, North America and Japan, other, less visible pollutants have come to predominate in urban areas. These are largely the result of combustion in vehicle engines (petrol and diesel) – oxides of nitrogen (NO_x) and VOCs – and the products of photochemistry – ozone and secondary aerosols (particulates) (see Chapter 4). Different patterns of economic and social activity have also resulted in changes in the seasonal and diurnal pattern of pollution.

Over time it appears that levels of urban air pollution first increase and then decrease as wealth increases and abatement technologies can be applied. Although the income at which this switch over in air pollution levels occurs is variable, it has been estimated that in almost all cases it is before per capita incomes reach US$ 8,000 (Fenger, 1999). In this context rising standards of living and technological development have been seen as necessary prerequisites for better environmental conditions (The World Commission on Environment and Development, 1987). High standards of living and access to advanced technologies do not, however, preclude poor standards of urban air quality as some examples later in this chapter will illustrate.

Although there are isolated complaints relating to air pollution in cities throughout the historical period, it appears that concerns about urban air quality first came to general notice in medieval London as people switched from wood to coal as their major fuel source (Brimblecombe, 1987). In the late thirteenth century a commission was set up to investigate pollution from coal burning in London and a proclamation banning the burning of sea coal (brought by ship to London from the coalfields of north-east England) was introduced in 1306. It had little or no effect. In medieval London the production of lime in the summer was also a major source of smoke pollution, giving a different seasonal pattern from the winter smokiness of the eighteenth, nineteenth and early twentieth centuries. Air pollution resurfaced as an issue in the Tudor London of the sixteenth century as the price of wood rose and there was again a switch to coal (Brimblecombe, 1987). 1661 saw the publication of John Evelyn's *Fumifugium or The Inconvenience of the Aer and Smoak of London Dissipated*. Evelyn's book described the impacts of smoke pollution on buildings, human health and clothes, identified industry as the major source (suggesting that polluting industries should be moved out of the city) and recommended that coal should be replaced as a domestic fuel by wood, charcoal or coked coal. During the eighteenth and nineteenth centuries conditions got worse, both in London and in other cities such as Manchester, Newcastle and Glasgow. Indeed, conditions in the newly industrialized cities were probably worse than in London. There were no organized efforts at smoke abatement until the nineteenth century, and no act was passed until 1853 (for London only) and further legislation came through public health and other sanitary legislation (see Section 7.1). The industrial lobby fought all attempts to tighten legislation, although the role of low level sources was increasingly recognized.

By the nineteenth century London's air pollution problems were being compounded by increasingly severe fogs, when periods of atmospheric stability trapped pollutants close to the ground. The thick, yellow/orange fogs caused severe disruption to normal life and were associated with higher than normal mortality rates. A week of fog in 1873 was associated with 700 excess deaths (Brimblecombe, 1987). The term 'smog' for the mix of smoke and fog was coined in 1905 (by Des Voeux). Smog became synonymous with London and no detective novel or film of Victorian London was complete without one! Smoke control legislation was slowly introduced during the early twentieth century and London, Manchester and Coventry took on powers to set up smoke-less zones. Coal continued to be the dominant energy source, however, with nearly 200 million tonnes of coal being used in 1960. The great London smog of December 1952 (see Section 5.5) marked another turning point, resulting in the passage of the 1956 Clean Air Act (see Chapter 7).

The availability of alternative fuels (e.g. gas, electricity), changing lifestyles and the changing nature of industrial activity made many UK cities visibly less polluted during the latter part of the twentieth century. The most dramatic change in all UK cities over the second half of the twentieth century was the steep decline in measured concentrations of smoke and SO_2 (Eggleston et al., 1992). In fact, a limited number of measurements of soot and sulphate in rain made in the late nineteenth century suggest that concentrations of these pollutants had already declined by the late 1950s (Brimblecombe, 1987). Big reductions in SO_2 concentrations, however, do not seem to have occurred until the 1960s. An analysis of SO_2 measurements made at the site of County Hall on the south bank of the Thames in London

(Laxen and Thompson, 1987) showed no clear trend from 1931 until about 1963/64 when a steady decline began. Trends in smoke (black smoke) and SO_2 concentrations at Stepney in London are shown in Fig. 5.1 (to convert $\mu g\ m^{-3}$ to ppb, divide by 2.704).

The pattern of SO_2 concentration in London has closely followed changes in emissions, which fell from some 485,000 tonnes in 1962 to 43,000 tonnes in 1988. In the late 1960s and early 1970s most of the decline could be attributed to a reduction in the use of solid fuel and over the latter part of the period to a decline in the use of fuel oil (Eggleston et al., 1992). Over the UK as a whole, total coal consumption had dropped to 58.8 million tonnes by 2002, while domestic coal consumption fell from 36 million tonnes in 1960 to 1.8 million tonnes in 2002 (DUKES, 2003). Consequently, few urban areas in the UK now experience persistent SO_2 problems with an average concentration of 7.7 ppb (20.6 $\mu g\ m^{-3}$) in urban areas in 1997–98. In fact, there is often little difference now between SO_2 levels in urban areas and the surrounding countryside (QUARG, 1993). Although it should be noted that short periods with high concentrations can

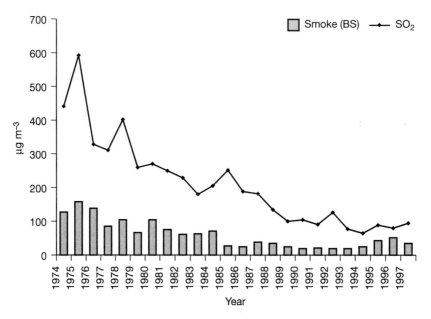

FIGURE 5.1 98th percentile SO_2 and black smoke (in $\mu g\ m^{-3}$) at Stepney, London (Data from the UK Air Quality Archive)

occur as a result of the grounding of the plumes from large industrial sources such as power stations. Seasonal patterns of SO_2 have also changed with a more even distribution between winter and summer replacing the previous winter peak. One city which did not follow the general trend was Belfast in Northern Ireland which had no access to natural gas to substitute for coal or smokeless fuel (QUARG, 1993). This is now changing and concentrations in Belfast are declining. Further reductions in the S content of liquid fuels mean that SO_2 concentrations are likely to continue their downward trend. Concentrations of black smoke have also continued to decline, with high levels confined to areas such as south and west Yorkshire and parts of Northern Ireland where coal is still a major source of domestic heating. The focus of concern in urban areas in the UK has moved towards NO_x, VOCs and particulates as primary pollutants and ozone and particulates as secondary pollutants, all mainly related to emissions from motor vehicles.

Information about air quality in London can be obtained at the following websites: http://www.airquality.co.uk and http://www.londonair.org.uk/etc./

Urban air quality concerns followed a similar pattern in the USA. The urban population (in towns of more than 2500) increased from about 6 per cent in 1800 to about 20 per cent in 1860 and 51 per cent in 1920. Between 1870 and 1920 the industrial city, relying on coal as its main source of energy, was the dominant urban form in the USA (Melosi, 1980). Areas which relied on soft, bituminous coal (with a high S content) were particularly badly affected. Pittsburg, which in 1904 produced 64 per cent of the USA's pig iron and > 53 per cent of its steel, fell into this category and was called 'the smoky city'. Until the 1920s, industrial activity was concentrated in the north-east states of the USA and these areas experienced the worst air pollution. By the late nineteenth century bad smog episodes (referred to as 'Londoners') were being recorded and, as coal use expanded still further in the 1880s and 1890s, so demands for smoke abatement increased. St Louis (which used bituminous coal) drafted an ordinance against smoke in 1893

although it was later ruled unconstitutional by the state legislature (Grinder, 1980). St Louis was subsequently granted the right to impose such ordinances in 1901.

By 1912 nearly every major city had a smoke inspector and at least some form of (weak) legislation. Anti-smoke legislation was, however, halted by the outbreak of World War I and the 1920s saw fewer complaints about smoke and the almost complete abandonment of the idea of prosecuting offenders. During the 1930s and 1940s smoke became less and less of an issue in the cities of the USA, but more as a result of new technologies than of tighter legislative controls. The early twentieth century also saw the rise of the car. In 1903 there were fewer than 10,000 cars in the USA, by the 1920s there were about 26 million, by 1980 there were 110 million and by 1999 the figure had reached 215 million. The car transformed the city in the USA as people moved from the city centres to the suburbs and car journeys became part of almost all activities. New forms of urban air pollution also came to the fore.

Many of the air pollution problems faced by cities in the developed world in the nineteenth and early twentieth centuries are now being experienced by the growing cities of the developing world. In addition to the 'traditional' pollutants from coal (and then oil) combustion, these cities are having to deal with the pollutants which result from vehicle emissions. The case studies below attempt to represent the range of conditions affecting urban populations under different circumstances of economic development and history of pollution control.

5.4 Case studies

Some of the issues relating to urban air quality are discussed below using three case studies from metropolitan areas with significant, yet different, air pollution problems. The cities described are Los Angeles on the west coast of California in the USA, Mexico City in the central highlands of Mexico and Beijing in eastern China.

5.4.1 Los Angeles and the South Coast Air Basin

'In terms of air pollution, there are probably few areas less suited for urban development' (CARB, 1999). This rather discouraging description was applied to the area of the South Coast Air Basin in California, home to the City of Los Angeles and about half the state's total population, by the California Air Resources Board (http://www.arb.ca.gov). Los Angeles (El Pueblo de Nuestra Senora la Reina de los Angeles del Rio de Porciuncula) was founded in 1781 as a result of an expedition sent out by the Inspector General of New Spain (later Mexico) to found new mission colonies in California. The new settlement was designed for 24 families, but only 12 families could be persuaded to move there from Sinaloa (The Smithsonian, 1989). The basin was not a promising location, isolated by the mountains, with a poor harbour and scarce water supplies. By 1870, Los Angeles' population had only reached 5000, but the railway boom of the 1880s saw this increase to about 100,000 by 1900. Other major periods of population growth occurred from 1900 to 1914 and in the 1920s, so that by 1930 the population had reached about 1,250,000. Between 1960 and 1990 the population of the Los Angeles Basin rose by 81 per cent and currently stands at about 14 million people and 9 million vehicles. It has been estimated that some 70 per cent of the surface area of Los Angeles is devoted to vehicles in the form of roads, car parks, petrol stations etc. There is little heavy industry in the Los Angeles Basin, but the area experiences poor air quality associated with vehicle emissions and photochemical pollution generated by long hours of sunshine and frequent stagnant air conditions. Offshore lies the cool California current which stabilizes the atmosphere and generates onshore breezes during the day carrying pollutants further inland where they are then trapped by the mountains.

Photochemical pollution began to be noted in the 1940s when it was associated with extensive crop damage. Los Angeles had its first recognized smog episode in 1943. This took the form of a photochemical haze, with high concentrations of O_3 (see Chapter 4), and the mechanism of its formation was identified in the early 1950s by Arie Haagen-Smit who was working in Los Angeles. The City of Los Angeles instigated its air pollution control programme in 1945 with the setting up of a Bureau of Smoke Control. In 1947, the Los Angeles Air Pollution Control District was established under California's Air Pollution Control Act; it was the first of its kind in the USA. Reliable ozone measurements began in 1965 when maximum hourly concentrations regularly exceeded 500 ppb (1000 $\mu g\ m^{-3}$). The first emissions control technology for private passenger vehicles became a requirement in 1963 and the first exhaust standards for CO and NM-VOCs were set in 1966.

As the overall situation in California deteriorated, the California Air Resources Board was created in 1967 with a series of air pollution control districts. The state was divided into 15 air basins, based on their geographical and meteorological coherence. The South Coast Air Quality Management District (SCAQMD), which includes Los Angeles, Orange and Riverside counties and the non-desert part of San Bernadino County, is the largest of the air basins (Lloyd, 1997). Information about this area can be found at http://www.aqmd.gov/aqmd and through the US EPA site http://www.epa.gov/region09/air/index.html).

By the time of the Federal Clean Air Act of 1970 (see Chapter 7) it was already recognized that California had particular problems and the state was allowed to introduce more stringent emissions standards. Catalytic convertors were introduced in California in 1974 and the quantity of Pb in petrol was in any case restricted by the CARB from 1976. Testing of vehicle emissions to ensure compliance with standards began in 1980. The California Clean Air Act was passed in 1988 and resulted in new regulations for cleaner vehicles and fuels. A low-emission vehicle programme was adopted in 1990 which aimed for 10 per cent zero-emission vehicles by 2003. A range of new, cleaner fuels, with new emission standards was also introduced. The early 1990s saw a shift towards the use of market mechanisms (trading emission credits) to control and reduce emissions. However, an economic downturn across California in the early

1990s changed political attitudes towards pollution control and it was suggested that the state's stringent environmental regulations might have contributed to its economic problems. The state legislature cut the SCAQMD budget and its powers and 9 of SCAQMD's 11 scientific advisors resigned in protest (Lloyd, 1997).

Over the period 1982–2001, California's population increased by 40 per cent and vehicle miles travelled by 97 per cent, but at the same time peak 1-hr O_3 concentrations fell 55 per cent and peak CO levels by 54 per cent (CARB, 2003). Between 1988 and 1997, PM_{10} concentrations fell by 31 per cent. Emissions of NO_x, VOC and CO have also fallen, mainly due to controls on petrol vehicles, but on-road vehicles remain the major source of all these pollutants. Emissions and concentrations of Pb fell dramatically as catalysts required the use of unleaded petrol. Emissions of SO_2 fell about 40 per cent between 1985 and 1995, mainly as a result of fuel switching in stationary sources, but also due to the introduction of fuels with a lower S content. Unlike other areas of the USA, SO_2 is not a problem in California (Doyle, 1997).

In the South Coast Air Basin area, overall trends have been similar to California as a whole. NO_x emissions from on-road sources increased from 1985 to 1990 and then declined; emissions from other mobile sources (e.g. off-road vehicles) increased. The South Coast Air Basin was one of the few areas in California to have a continuing NO_2 problem and was designated a non-attainment area with respect to federal air quality standards. In the early 1990s, a number of sites breached the WHO 1-hr NO_2 guideline. The latest data, however, suggest that both state and federal NO_2 standards are being met. Emissions of CO have also shown a pattern of decline from on-road vehicles and an increase from off-road vehicles. State and federal standards were still exceeded in 1997, but only in small areas of Los Angeles county and by 2001 were attained in most parts of California. VOC emissions fell from 1985 onwards, even though the number of vehicle miles travelled per day increased 75 per cent between 1980 and 1997 (CARB, 1999). This was due to strict emissions controls on petrol vehicles. Across the state as a whole, emissions per

vehicle of NO_x fell by 58 per cent between 1970 and 1995, and emissions of VOCs by 80 per cent.

In spite of reductions in ozone precursors, ozone continues to be a problem in California in general and in the SCAQMD in particular. Over the period 1995-97, California experienced two of the ten maximum hourly O_3 concentrations recorded across the USA. Both of these were in the SCAQMD area. This still represented an improvement; in the 1960s concentrations in excess of 600 ppb were not uncommon, by the end of the twentieth century peaks were around 200 ppb.

The pattern of exceedances of the federal 1-hr standard for 2000 is illustrated in Fig. 5.2a. The central San Bernadino Mountains had the highest number of days of anywhere in the USA when the federal ozone standard was exceeded.

Figure 5.2a illustrates that the highest exceedances (and therefore ozone levels) occur east of the centre of the city of Los Angeles. This reflects the balance between NO_x and VOC emissions, with the greatest production of ozone taking place away from high NO emission areas. Trends in maximum ozone concentrations and exceedances of state and federal standards (90 ppb and 120 ppb respectively) in the SCAQMD are illustrated in Fig. 5.3.

There is massive exceedance of the WHO standard.

1997 saw only one stage I episode (concentration above 200 ppb) compared with 121 in 1977. Stage I episodes (smog alerts) represent periods when everyone is advised to avoid strenuous outdoor activity and susceptible groups (i.e. those with lung or heart problems) are recommended to stay indoors. Although the general trend is undoubtedly down, it has been recognized that the El Niño of 1997, which resulted in lower temperatures and increased cloudiness, reduced ozone formation (South Coast Air Management District, 1997). Ozone levels were higher in 1998, but the overall trend was still down. In 1999 and 2000, there were no stage I episodes in the South Coast Air Basin.

Concentrations of PM_{10} have continued to exceed state and federal standards, although concentrations have declined since 1988. Exceedances of the federal standard occur in only a limited area (Fig. 5.2b), but about 50 per

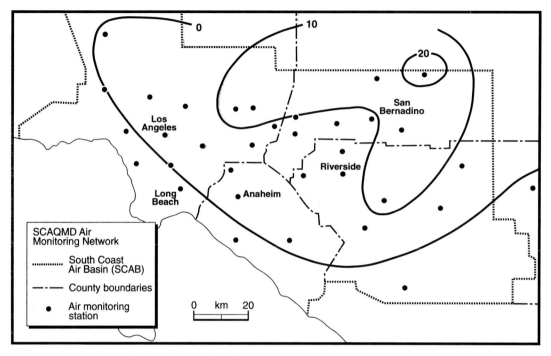

FIGURE 5.2A Pattern of exceedances of the federal 1-hr standard for ozone (120 ppb) over the SCAQMD in 2000 (Source: SCAQMD Current air quality and trends, 2000)

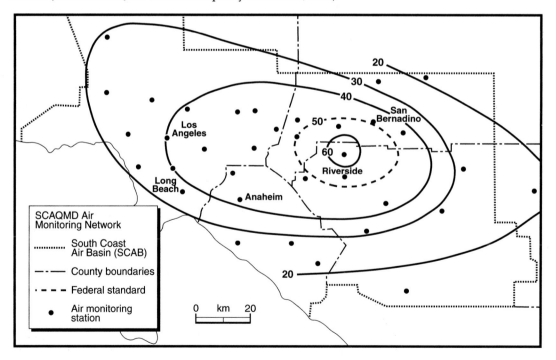

FIGURE 5.2B Pattern of exceedances of the federal 1-hr standard for PM_{10} (50 $\mu g\ m^{-3}$) over the SCAQMD in 2000 (Source: SCAQMD Current air quality and trends, 2000)

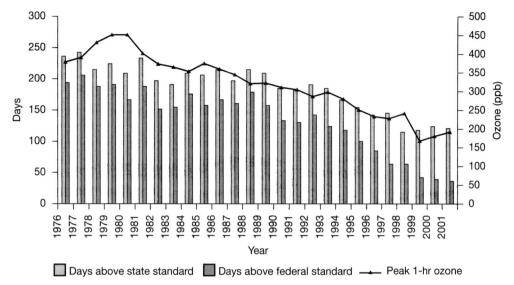

FIGURE 5.3 Trends in maximum ozone and in exceedances of federal and state standards in the SCAQMD, 1976–2001 (Data from http://www.aqmd.gov/smog/°3trend.html)

cent of the South Coast Air Basin exceeded the state standard in 1998. Across California as a whole most areas exceed either the state 24-hr or annual PM_{10} standard at some time. The more limited data available for $PM_{2.5}$ show substantial exceedance of the state $PM_{2.5}$ standard, especially in the South Coast and San Joaquin Valley air basins (CARB, 2003).

Total direct emissions of primary particulates have not changed significantly. Some parts of the SCAQMD experience high levels of PM_{10} when windstorms bring in dust from adjacent desert areas.

Los Angeles has constantly struggled to meet federal air quality standards for pollutants associated with vehicle emissions, in spite of extremely stringent emission control standards. Ozone has been the main problem for some 50 years. The 1990 Federal Clean Air Act Amendments (see Chapter 7) gave the city a further 20 years to meet the federal standard. The South Coast Air Basin is the only extreme non-attainment area for ozone in the whole of the USA, meaning that it has been recognized as a difficult problem which will take a long time to address successfully. A recent development has been concern over the role of fuel additives in driving ozone production. In 1999, California

agreed to ban an organic additive which is widely used to raise the octane level of unleaded fuel, MTBE (methyl tertiary-butyl ether). This will be phased out by 2004. An Air Quality Management Plan was introduced in 1997 to address the continuing problems with ozone, CO and PM_{10}, but by the end of 1999 this plan had received only partial approval from the EPA.

5.4.2 Mexico City

When the Spanish, led by Cortes, entered the basin of Mexico from the east in 1519, they found a basin that was already densely populated. The Mexica (Aztec) city of Tenochtitlán, later the site of Mexico City, had an estimated population of 200,000 – 300,000. At the time, Cortes marvelled at the wonderful views across the basin with the lakes on the basin floor and the mountains around it culminating in the snow-capped peaks of Popocatépetl and Iztaccihuatl (both > 5000 m). Today it is rare to be able to see these volcanoes from Mexico City, especially in the dry season, due to the persistent haze of polluted air which hangs over the city. Mexico City now has a population in excess of 20 million, more than 18 per cent of the national total (Lacy, 1993). The main city (Distrito Federal – DF), together with 17 municipalities from the Estado de Mexico, form the Zona Metropolitana de la Ciudad de Mexico (ZMCM), the urban conurbation with a built-up area of about 1500 km^2. Like Los Angeles, Mexico City is not in a good location in terms of maintaining good air quality. It lies some 2240 m a.s.l. at about 19°N and is surrounded by mountains – atmospheric inversions are common and low oxygen concentrations slow oxidation. Rainfall is highly seasonal, falling mainly between June and September. Local winds tend to blow from the north across the city centre towards the south-west. Some background material relating to the issues affecting Mexico City can be found at http://www.ess.co.at/GAIA/CASES/MEX/index.html.

The population of Mexico City was estimated at about 100,000 in the mid-eighteenth century (less than at the time of the Spanish conquest). It underwent a major period of growth between 1880 and 1910 when the built-up area expanded to five times its previous size; the National Census of 1900 recorded a population of 345,000. The city underwent a major phase of industrialization and growth from the 1950s to the 1970s with no attempts to control emissions. By 1990, the population of the metropolitain area was estimated at 15 million (8.2 million in the DF) and by the late 1990s at 20 – 22 million. In spite of economic crises in the 1980s, emissions continued to increase. The number of vehicles increased from fewer than 50,000 in 1940, to 680,000 in 1970 and 2.7 million in 1993. Private cars comprise some 71 per cent of the total number of vehicles. It has been estimated that there are some 145,800 taxis, about 480,000 lorries and 51,000 buses and minibuses in the ZMCM (INEGI, 1998). Many of these vehicles are quite old (the median age of the vehicle fleet was estimated at 12 years in the mid-1990s) and/or poorly maintained so that emissions per vehicle are much higher than in the USA (Streit and Guzman, 1996). From mid-1988 until the end of 1991 alone, petrol sales in the Basin of Mexico rose by 31 per cent. Unleaded petrol (Magna Sin) was only introduced in 1990 and

initially was not widely available. The take up was also relatively slow due to its cost and the predominance of older vehicles.

The ZMCM is a major industrial centre. At the start of the 1990s it had more than 44,000 industries of which 72 per cent were in the DF itself, mainly located in the north-east part of the city. Most of these were small enterprises, but major polluting industries included two large cement plants, a PEMEX refinery at Azcapotzalco (closed in 1991), two power stations (now switched to natural gas) and about 60 large chemical, paper and steel plants. Industrial sources appear to have been relatively minor contributors to overall pollution loads, but they have been the primary sources of SO_2 and particulates. There are also problems with primary particulates derived from soil and dry lake sediments, especially in the north and east of the city.

Air pollution monitoring began in Mexico City in the 1950s and in 1966 a network of 14 sites was installed to measure smoke, TSP (using high volume samplers) and SO_2. During the early 1970s, with the assistance of UNEP, a manual network for TSP and SO_2 was installed, which continues to operate. There are 19 sites in the manual network. The focus is on particulates (Total Suspended Particulates, PM_{10}, $PM_{2.5}$) with NO_3, SO_2 and heavy metals. An automatic monitoring network (RAMA) for SO_2, CO, O_3, NO, NO_2, NO_x and H_2S came into operation in 1985, having been set up with assistance from the US EPA. PM_{10} and $PM_{2.5}$ are now measured as well. The number of sites in RAMA has recently been increased to 36 from 32 (http://www.sma.df.gob.mx/simat/pnrama2.htm). The monitoring of VOCs was expected to begin in 2004 (GDF, 2003), but no data had been reported by the end of the year.

A coordinating system for air quality and acid deposition measurements, SIMAT (Sistema de Monitoreo Atmosferico de la Ciudad de Mexico), started in 2000. Air quality standards are expressed in two ways: limits for individual pollutants and an overall index of air quality (IMECA – Indice Metropolitano de la Calidad del Aire). Air quality standards are set out in Table 5.3.

The IMECA index is an indication of overall air quality and is used to provide information

TABLE 5.3 Air quality standards in Mexico (Data from http://www.sima.org.mx/valle_de_mexico/mexico.htm)

Pollutant	Standard	Averaging time
PM_{10}	150 µg m^{-3}	24 hrs
TSP	260 µg m^{-3}	24 hrs
CO	11 ppm (11,000 ppb)	8 hrs
O_3	110 ppb	1 hr maximum
NO_2	210 ppb	1 hr maximum
SO_2	130 ppb	24 hrs
Pb	1.5 µg m^{-3}	3 month mean

Note: no air quality standards have been set for VOCs.

to the general public. The scale runs from 0 to 500 in four bands (Table 5.4).

Within each band, threshold concentrations of TSP, SO_2, NO_2, CO and ozone are set. The air quality standards in Table 5.3 correspond to 100 on the IMECA scale. There is also a global IMECA index with broad categories: 0 – 100 = satisfactory; 101 – 150 and 151 – 200 = not satisfactory; 201 – 300 = bad.

During the 1970s and 1980s there was a general increase in awareness of environmental issues in Mexico, with national laws for the environment in 1972, 1981 and 1988. The first separate environmental body, SEDUE (Secretaria de Desarollo Urbano y Ecologia), was created in 1981 and included a directorate to control and prevent pollution. The directorate had powers to inspect factories and prosecute those found breaching national standards, and they also managed the monitoring systems.

SEDUE was later replaced by SEDESOL. Action was also taken within the DF itself with the creation, in 1983, of a subdirectorate for the prevention and control of air pollution. An integrated programme to deal with air pollution (PICCA) in the ZMCM was introduced in 1990 and replaced in 1996 by a new scheme (PROAIRE) with additional powers. The latest initiative under PROAIRE is the Programme to improve air quality in the Metropolitan Zone of the Valley of Mexico 2002–10.

By the mid-1980s, the seriousness of air pollution in Mexico City led to the first concerted attempts to reduce emissions. Gas replaced coal for some power generation, highly polluting industries were closed or relocated, a programme of vehicle emissions testing was introduced and the lead content of petrol was reduced. A programme to restrict car usage – 'hoy no circular' (can't drive today) – was started in 1989, initially for the winter months only, but then all year round. All new petrol vehicles were to be fitted with catalysts from 1991. In spite of these activities, the UNEP/WHO report of 1992 recorded that the ZMCM had serious air pollution problems and that it was getting rapidly worse. WHO standards for SO_2, SPM, Pb and NO_2 were all being exceeded on a regular basis. Ozone levels were noted as being exceptionally high, particularly in the south-west of the city where 60–80 per cent of days exceeded the national standard; the WHO guidelines are exceeded by more than a factor of two. The introduction of PICCA was noted as an important step, but one whose successful implementation would require a great deal of effort.

Over the 1990s, there has been some improvement in air quality, particularly for SO_2 and Pb. Between 1993 and 2002, for example,

TABLE 5.4 IMECA (Indice Metropolitano de la Calidad del Aire) index of air quality

0–100	Satisfactory	Suitable for all types of activity.
101–200	Unsatisfactory	Minor problems for sensitive groups.
201–300	Bad	Increasing problems and intolerance to exercise for people with respiratory conditions.
301–500	Very bad	Various symptoms and intolerance to exercise across the population.

there were no days which exceeded 300 points on the IMECA scale (Instituto Nacional de Ecologia, 2003). However, ozone exceeds the standard (110 ppb) for most of the year, with the highest levels in the south-west of the city and the lowest in the north-east (Fig. 5.4).

This spatial distribution reflects the prevailing wind direction and the evolution of ozone downwind of the city centre (as in Los Angeles, see above). Even for ozone, however, there has been a reduction in peak concentrations (Fig. 5.5).

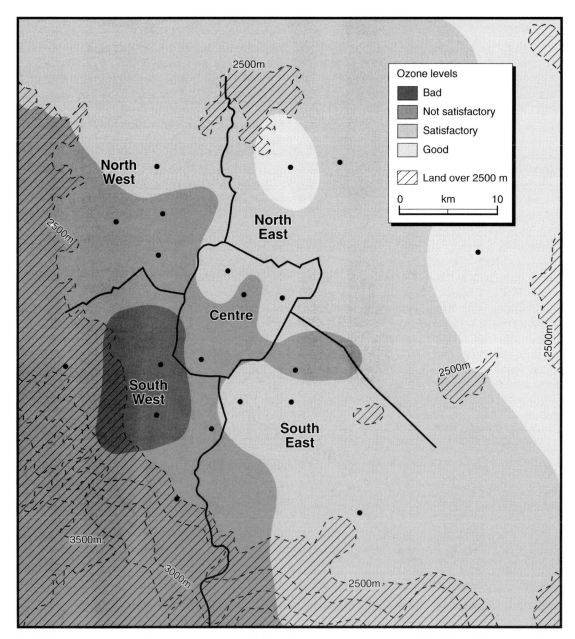

FIGURE 5.4 Spatial distribution of ozone in Mexico City and its conurbation (ZMCM), 1999 – using the IMECA categories (Data from http://www.sma.gob.mx/imecaweb/mapas/diptico.pdf)

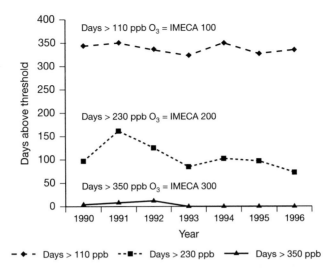

FIGURE 5.5 Days with ozone levels above the IMECA standards in the ZMCM (Data from INEGI, 1998)

There are few exceedances of the 110 ppb standard in the cool, dry winter months, but between March and May (warm and dry) in 2001–02, the standard was still exceeded for nine out of ten days (GDF, 2003).

As major industrial pollution sources have been controlled, or closed, so vehicle emissions have become more and more dominant in the Basin of Mexico. Transport dominates emissions of VOCs (52.5 per cent), NO_x (75.4 per cent) and CO (96.7 per cent) (Lacy, 1993). The adoption of a low leaded petrol in the late 1980s resulted in the expected decline in Pb concentrations in air, but had the unexpected impact of increasing ozone levels. The reason for this was the addition of high alkyl isomers and alkyl aromatic fractions to the new fuel which produced highly reactive VOCs. When Magna Sin (unleaded fuel) was introduced in 1990 it was reformulated (including MTBE – see discussion re California above), but ozone concentrations, after declining in the early 1990s, then rose again. Pb concentrations fell from more than 2.5 μg m^{-3} in 1986 to 0.8 μg m^{-3} in 1993. Consumption of unleaded fuel in the ZMCM only exceeded that of leaded petrol from 1996 onwards (Bravo and Torres, 2000). Other efforts to control emissions have included the extension of the city's metro system into some of the outer suburbs, the replacement of old buses with low emission vehicles and a programme to convert passenger and goods vehi-

cles to run on LPG (liquid petroleum gas). The state oil company, PEMEX, has also promised to deliver an 'Ecological Package' including both new fuels (e.g. low S diesel) and improved vapour recovery systems for storage tanks and petrol stations.

Ozone is now *the* pollution problem in Mexico City. In the UNEP/WHO report of 1992, Mexico City was the only megacity with worse ozone pollution than Los Angeles. Mexico's own hourly ozone standard was exceeded on 71 per cent of days in 1986 and 98 per cent of days in 1992. The failure to control Mexico City's ozone problem can be attributed to the increasing number of vehicles (many people bought additional, older cars in response to the 'hoy no circular' restrictions) and the fact that, even by 1997, 65 per cent of the petrol car fleet were still pre-1991 no-catalyst models. The lastest vehicles (1997–99) still have greater ozone creation potential than their US equivalents. From 2001, however, new Mexican cars have had to meet exactly the same standards as those imposed on US vehicles, which should reduce emissions of both NO_x and VOCs.

Although vehicles are likely to remain the primary target for emissions controls, it has also been suggested that LPG (widely used in the city for cooking and heating and increasingly in vehicles) could also be an important source of ozone precursors (Blake and Sherwood Rowland, 1995). LPG for domestic use often leaks from storage cylinders and contains high concentrations of highly reactive C_3 and C_4 alkanes. This composition of LPG is different from that in use in the USA, where the hydrocarbons are mainly propane. Improved storage and a change in composition could make a useful contribution to reducing O_3 levels. Particulates are also a serious problem, particularly in the east of the city where dust is a major component. Between 40 and 60 per cent of TSP are PM_{10}. Daily TSP concentrations exceed the Mexican standard of 275 μg m^{-3} more than 90 per cent of the time. The old WHO TSP annual mean guideline of 60–90 μg m^{-3} was exceeded at all the city's monitoring sites through to 1992. As more than 90 per cent of particulates affecting Mexico City may come from 'natural' sources, such as erosion and forest fires, controlling their levels will be highly problematic. The paving of roads and pavements and erosion control measures such as tree planting may help to some extent. It has been noted, however, that particulate loadings in Mexico are no worse than they were in Los Angeles in the early 1970s.

As one of the world's 'megacities' Mexico has become a focus for air quality studies and there have been a number of international programmes to collect high quality data. Information on one of these, based at the Massachusetts Institute of Technology, can be found at http://www-eaps.mit.edu/megacities/overview/program.html. Molina and Molina (2002) have provided an overview of the air pollution history of the ZMCM and the challenges it still faces.

5.4.3 Beijing

Beijing was first declared a capital city in 1057 BC and in modern history has been the capital of China since the thirteenth century. It lies on the border of the Great North China Plain, with the Mongolian plateau to the north and mountains to the north-east. There is a small, densely populated old walled city surrounded by wide reaching suburbs. The city had a population of about 12 million in the late 1990s and has been identified as one of the world's worst ten cities in terms of its air pollution. Its population has grown from 8.3 million in 1970 (UNEP/WHO, 1992) which means a slower rate of growth than other major cities in developing countries, probably due to the restrictions on travel within China during the Cultural Revolution and the recent past. Urban air quality is generally poor across China with 500 major cities not meeting the WHO guidelines. Beijing, Shenyang and Xian are three of the top ten most polluted cities in the world (WWICS, 1997).

China has set standards for ambient air quality for three classes of land use: I – tourist, historical and conservation (the most stringent); II – residential urban and rural areas and III – industrial and areas of heavy traffic (least stringent). Some of these are given in Table 5.5.

Coal has traditionally been the main source of energy with Beijing alone using 21 million tonnes per year (the UK's total consumption in 1998 was 62.9 million tonnes). Chinese coal generally has a high S content. Urban use of coal is high and it has been estimated that this type of usage accounts for 75 per cent of the total output of pollutants in China (WWICS, 1996), including 18.25 million tonnes of SO_2 (the UK total SO_2 emission for 1998 was 1.6 million tonnes). As well as being the major political and cultural centre, Beijing is also a major industrial city with nearly 6000 industrial enterprises including 24 power plants, 53 metal smelters, 18 coking plants and 194 chemical plants. An emission inventory for Beijing prepared by Krupnick and Sebastian (1990) indicated that its total annual SO_2 emissions were about 530,000 tonnes, with industry contributing 36 per cent, power generation and coking 24 per cent and household stoves some 14 per cent. (The total should be regarded as a minimum, as a number of source categories were not included.)

SO_2 and particulates have been the traditional pollutants of industrial, urban areas of China. Results from monitoring in the centre of Beijing from 1980 to 1990 showed that SO_2 levels breached WHO guidelines. There was also a clear annual pattern to the concentrations, with the highest levels in winter (November to

TABLE 5.5 Selected air quality standards for China (Source: Sinton *et al.*, 1998)

Pollutant	Class	Daily	Maximum permissible
TSP (μg m^{-3})	I	150	300
	II	300	1000
	III	500	1500
SO$_2$ (μg m^{-3})	I	50	150
	II	150	500
	III	250	700
CO (μg m^{-3})	I and II	4	10
	III	6	20
NO$_2$ (μg m^{-3})	I	50	100
	II	100	150
	III	150	300

March) reflecting the need for domestic heating. Over the same time period, levels of TSP were extremely high (UNEP/WHO, 1992), although the city is regularly hit by dust storms originating in the plains. Even if half the measured TSP was assumed to come from dust storms, however, the remainder would still exceed the WHO guideline. Most of the anthropogenic TSP is believed to come from coal combustion, as local coal has a high ash content and the trace metal composition of the TSP would also suggest this source. The WHO guideline based on daily values was exceeded on almost every day of 1989 at a monitoring site to the west of the city near an iron and steel complex. The maximum recorded daily values were >1000 μg m^{-3}. A more recent study of PM$_{10}$ and PM$_{2.5}$ (Sun *et al.*, 2004) has confirmed higher concentrations of particles in winter than in summer, with a significant percentage coming from long range dust in the winter/spring. Efforts have been made to reduce pollution, especially from particulates and from 2000 coal was being replaced by gas for domestic heating. Coal of high ash and sulphur content was also prohibited. In spite of these steps, in 2001 the 150 μg m^{-3} national air standard was only met half the time (Shi *et al.*, 2003). Since 2000, each city in China has had to publish an air pollution index (scale 0 – 500).

At the time of the UNEP/WHO report, concentrations of pollutants associated with motor vehicles (lead, NO$_2$) were not felt to be a problem, although measurements were limited. CO levels appeared to be higher in winter than summer and much higher indoors than out. High indoor concentrations were associated with the use of coal for domestic cooking. Over the 1990s, however, the level of vehicle ownership has increased rapidly. There were only 710,000 private vehicles in China in 1991 and this had increased to 1,500,000 by 1995 (WWICS, 1997). By the end of 2003 there were >96 million vehicles, including 24 million cars. The changing nature of emissions is now indicated by the fact that Beijing, Shanghai, Lanzhou and Chengdu all now suffer photochemical smog. Monitoring data from Beijing (see Fig. 5.6) show that NO$_x$ and TSP are now the major pollutants – SO$_2$ only dominated for one recording period (February 2000) between March 1998 and June 2000. Data from 13 monitoring sites in the city can be accessed through the Internet (http://www.bje.gov.cn/english_homepage/index.asp). Information can also be found at http://www.zhb.gov.cn/english.

5.5 Episodes

The most extreme forms of urban air pollution are called episodes. These are periods of a few hours or days when very high concentrations of pollutants can be recorded. They are generally associated with the occurrence of unusually

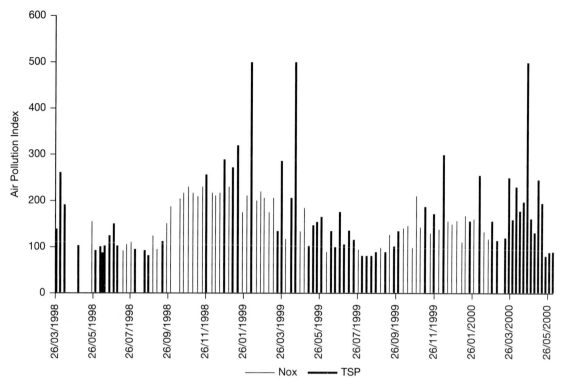

FIGURE 5.6 Air pollution index scores and dominant pollution in Beijing, 1998–2000

stable atmospheric conditions when high pressure predominates and wind speeds are low. Under such circumstances, pollutants accumulate and concentrations can reach very high levels. In common with the general temporal trends in urban air quality outlined above, so the nature of episodes has tended to change. Until the mid-twentieth century, most episodes occurred in winter and were associated with high levels of smoke and SO_2. More recently they have tended to be summer phenomena associated with high levels of NO_2 and particles in city centres and of photochemical pollutants, particularly ozone, in suburban areas. The winter episodes that occur also show high concentrations of pollutants such as NO_2, VOCs and particulates associated with vehicle emissions. Episodes are important not only because of the severity of their short-term effects (including health outcomes) but also their importance in prompting air pollution legislation (see Section 5.2). Some examples of winter and summer episodes in the UK are given below. Background information on air

pollution episodes can be found at http://www.airquality.co.uk frequently asked questions (faq), 'What is a pollution episode?'

Between 4 and 9 December 1952, London experienced a major traditional smog (smoke and fog) episode. Although the monitoring equipment was not very sophisticated, it showed that the daily mean smoke concentration reached 4460 µg m^{-3}, compared with a normal loading of about 250 µg m^{-3} (Brimblecombe, 1987). Daily average SO_2 concentrations were in the range 1000 to 1400 ppb (3000 to 4000 µg m^{-3}). This episode resulted in some 4000 additional deaths and played a key role in triggering the first Clean Air Act (see Chapter 7) and the establishment of a national smoke and SO_2 monitoring network (see Chapter 2). The most severe winter episode in London in recent years occurred between 12 and 15 December 1991. Weather conditions were again cold and foggy, with low wind speeds (< 3 m sec^{-1}), as a result of the influence of an anticyclone centred over the Alps (QUARG, 1993). In contrast to the 1952 episode, the major pollution peaks in December 1991 were associated with NO_2 and CO and could be associated with vehicle emissions. The maximum hourly concentration of NO_2 recorded was over 423 ppb, with three central London monitoring sites exceeding the EC hourly limit value of 105 ppb. In spite of this, there was no exceedence based on the 98th percentile, so there was no breach of the EC directive. Although NO_2 values in London were high during this episode, much higher values had been recorded in Glasgow in December 1981 (1180 ppb) and in London (Cromwell Road) in November 1982 (1820 ppb). A winter episode of lesser severity than December 1991, but broader geographical impact (from London to Belfast and Newcastle), occurred in December 1994.

In the UK, persistent periods of poor air quality in the summer are associated with unusually hot and sunny weather (cf. Los Angeles or Mexico City described above where such conditions occur routinely). High pressure, low wind speeds and abundant sunshine to drive photochemical reactions lead to the accumulation of pollutants such as NO_2 and ozone. The summer of 1995 was particularly hot and resulted in a number of air pollution episodes. Very high concentrations of ozone (>90 ppb) were recorded at many rural monitoring sites across the country on a number of occasions between late June and August. The site of Yarner Wood in south-west England recorded 130 ppb on 2 August. Although concentrations in urban areas are generally lower than in rural (where less ozone is consumed in the oxidation of NO), the site of London Bexley recorded 101 ppb on 1 August (PORG, 1997). Conditions suitable for a build-up of pollutants and the accumulation of ozone were the result of anticyclonic conditions dominating the country for much of the time. Major episodes were recorded 28 June–2 July, 30 July–3 August, 10–12 August and 17–22 August. Back trajectory analysis for 11 August showed that the air masses entering the UK had generally originated in the clean air of the Baltic/Gulf of Finland, before coming south into central Europe and then turning west and north to cross the UK. Although the values recorded in 1995 were high, they were still lower than those of the summer of 1976 when maximum concentrations in June and July exceeded 200 ppb even at County Hall in central London.

While in the areas around London summer episodes are increasingly typified by high levels of ozone, the city itself sees high concentrations of NO_2 and CO. Data from the monitoring site in Bloomsbury (central London) are shown in Fig. 5.7 (ozone values from Harwell to the west of London are shown for comparison).

The pollution episode of August 2003 has already been described in Chapter 4 in the context of ozone, but this episode also saw high PM_{10} concentrations (peak 24 hr mean = 62 $\mu g\ m^{-3}$ at Bloomsbury, London) (Stedman, 2004). It is estimated that 207 additional deaths were brought forward by the high PM_{10} levels over this period compared with the same period in 2002.

5.6 Indoor air pollution

The quality of air inside buildings is an aspect of air pollution which is often overlooked. This is perhaps surprising given that many people spend much more of their time indoors than out. It has been estimated, for example, that in the USA people spend 88 per cent of their time indoors, 7 per cent of their time in vehicles and only 5 per cent of their time outdoors (Jones, 1999). While this may be the extreme, the general pattern is typical of developed countries. Indoor air quality will be determined by a range of factors including outdoor air quality, the nature of indoor combustion (e.g. cooking, heating), smoking, levels of ventilation and materials used (e.g. paints, varnishes, insulating materials). Pollutants such as radon, asbestos, PAHs, mercury, a range of VOCs, aerosols and allergens which occur indoors will have predominantly indoor sources. Most indoor air pollutants have direct effects on the respiratory and cardiovascular systems, but the severity of these impacts will depend on a range of factors such as intensity and duration of exposure, the age and health status of the people exposed.

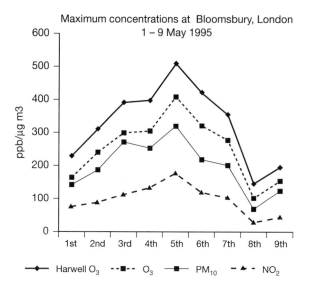

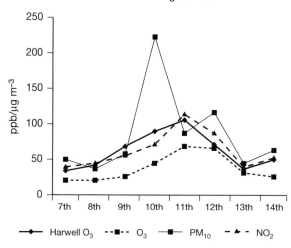

FIGURE 5.7 Pollutant concentrations during episodes in summer 1995 in the UK at Bloomsbury, central London – ozone values from Harwell, to the west of London, shown for comparison. Values in ppb except for PM_{10} in $\mu g\ m^{-3}$ (Data from the UK Air Quality Network)

The nature of indoor pollution tends to be different in developed and developing countries, with problems due to low ventilation rates (i.e. better insulation) in the former and due to heating and cooking in the latter (WHO, 2000). In developed countries, concentrations of pollutants inside are generally the same as outside except for short periods in areas such as kitchens or where there are badly maintained gas appliances. Information on current research in Europe can be found at http://wads.le.ac.uk/ieh/ierie/index.htm (Inventory of European Research on the Indoor Environment). In many developing countries, people can experience significant exposure to pollutants while indoors, particularly where solid fuels are used for cooking and heating (>75 per cent of households in China and India). High levels of particulate matter are a major problem, especially in rural areas where cooking often relies on the use of fuels such as wood in small, poorly ventilated spaces. A number of studies of TSP concentrations experienced by women in rural Indian villages cooking on wood, have shown concentrations in excess of 3600 μg m^{-3}. Biomass smoke contains a wide range of organic compounds (PAHs and VOCs) and the WHO has concluded that ' exposure to biomass smoke from open stoves causes considerable human ill-health world wide' (WHO, 2000). Where coal burning is important, in China and India for example, high indoor levels of fluorine and arsenic can also occur.

It has become apparent that indoor air pollution in China is a serious problem (Sinton *et al.*, 1998) with concentrations of SO$_2$ and particulates reaching the same levels as those recorded in the episodes in Donara (1948) and London (1952) (see above). Domestic energy use in rural areas is dominated by wood (34 per cent) and crop residues (38 per cent), while in urban areas coal predominates (56 per cent). A study of indoor air quality in non-smoking households in Beijing, using a mix of fuels, showed PM$_{10}$ concentrations ranging between 230 and 1067 μg m^{-3} in winter and 171 and 391 μg m^{-3} in summer. In a restaurant, where coal was the main fuel, a TSP concentration of 2920 μg m^{-3} was recorded. In rural areas, where biomass was the main fuel, measured indoor TSP concentrations were in the range 240 to 2570 μg m^{-3}, with an extreme PM$_{10}$ value in a room in Yunan province where wood was the cooking fuel of 22300 μg m^{-3}. These levels should be compared with the standards in Table 5.4 – the proposed PM$_{10}$ standard is about half the value for TSP. A number of Chinese studies have shown a statistically significant relationship between the use of coal for heating and cooking and the occurrence of lung cancer.

In all countries, smoking can be a major source of indoor air pollution resulting in high concentrations of CO, PM$_{10}$, nicotine and known carcinogens such as benzene and nitrosamines. Concentrations of PM$_{10}$ have been found to be two to three times higher in the houses of smokers than non-smokers. In places such as bars and nightclubs, extremely high levels of PM$_{10}$ have been recorded.

5.7 Conclusions

The nature of urban air pollution has clearly varied through both time and space. In most developed countries, pollutants which are largely invisible are now the major source of concern, while in developing countries the traditional visible pollutants, such as smoke, continue to be a major problem. The exact pollution climate experienced by any individual urban area may depend as much on its geographical context (particularly in terms of atmospheric stability) as on its level of absolute or per capita emission. Almost any urban area can experience air pollution episodes given the right combination of conditions, although these may be short lived. Conditions which give rise to poor air quality may be due to both local and regional emission sources.

In order to understand the nature of urban air quality, good monitoring is essential in terms of accuracy, spatial coverage and temporal resolution. International efforts provide limited support for such activities in the developing world. In attempts to relate levels of air pollution to health effects, there is clearly a need for much more monitoring of personal exposure. There has also been an increasing interest in developing detailed emissions inventories for urban areas to help in the formulation of policy, based on both measurements and air pollution models.

It is clear that traffic is a (if not *the*) major source of urban air quality problems around the world, as motor vehicles of all sorts become more and more numerous. Considerable efforts have gone into making individual vehicles, and the fuels they use, less polluting. Making this technology more widely available is obviously important. It is increasingly clear, however, that traffic management and ultimately traffic reduction is the only long-term solution. This is not always a popular policy option impinging as it does on areas such as personal aspiration and choice.

The whole area of indoor air pollution, at home, at work or actually inside vehicles remains under-researched.

6

Modelling atmospheric pollution and environmental change

6.0 Introduction

While it is certainly true that all our understanding of the environment is based on observations, it is not the case that observations can answer all of our questions about the environment. Many of the questions asked by policy makers go well beyond the scope of observations and address such issues as: What are the major sources of air pollution? Which are the most important sources to control? What levels of pollutant emissions are sustainable in the long term? Many of these questions require a combination of observations and mathematical modelling to answer. Modelling tells us whether we know enough about the environment to answer the policy makers' questions and to supply some of the possible solutions.

There are almost as many air pollution models as there are air pollution problems and it would be well beyond the scope of this chapter to provide a comprehensive review of them all. Instead, we set out some of the broad issues that underpin the approaches taken to air quality modelling and focus on a number of air pollution problems that cover spatial scales ranging from the urban to the global.

A numerical model comprises a set of relationships which link input parameters to output variables, causes to effects. They vary enormously in complexity and, inevitably, almost all are computer-based. Simple models can be represented as spreadsheets, whereas complex models may contain thousands upon thousands of lines of coding and may take months of execution time on the fastest supercomputers.

In principle, the values of the model output variables depend upon the chosen values of all the model input parameters. Sensitivity studies are used to define how the values of the output variables change in response to changing assumptions about the input parameters. Some model input parameters will be well defined because they are fundamental constants or because they are firmly based on observations. Other model input parameters may be uncertain because they are difficult to observe or because they represent gross simplifications of complex environmental processes which themselves are not well understood. Whatever the level of understanding is, best estimates of each input parameter should always be used so that bias in the model output variables is minimized.

Model testing and verification are essential steps in model building. Conversely, without adequate testing and verification against observations, modelling is likely to be meaningless. For policy makers to have some confidence in the model results and the answers to their questions, the model should be able to reproduce real world behaviour. The policy maker might reasonably ask if the model can simulate the concentration gradients across a monitoring network or the hourly and seasonal variations in concentration at a particular location. Sometimes, policy makers ask for model evaluation and model validation studies to be performed in addition to model verification studies. All these additional tasks are concerned with building adequate confidence into the modelling process so that policy makers can understand whether model statements about the outcomes of particular policy measures are likely to be reliable and robust.

In the sections that follow, many of the underlying issues in the design, formulation and application of models are illustrated by focusing, in turn, on urban, regional and global air quality.

6.1 Urban air quality

6.1.1 Urban air quality management

During the last century, more and more of the world's population moved into urban areas to live and work (see Chapter 5). Concerns about urban air pollution and human health effects have grown rather than declined, despite significant progress being made in a number of areas: reducing motor vehicle emissions; the design of more efficient heating and cooking systems in homes; tackling the gross pollution problems of industries and the substitution of clean fuels and energy systems.

Urban air quality management describes the process by which policy makers set about improving the quality of life in urban environments by reducing the impacts of air pollution on human health. Essential elements in the urban air quality management process include:

- An urban air quality monitoring network to establish the current levels of each of the major air pollutants.
- A set of air quality guidelines or standards against which current air quality standards may be judged and to act as targets to be reached through future policy actions.
- An emission inventory for each of the major air pollutants, covering all the major current pollutant-generating sectors and activities.
- A set of possible pollution control measures, together with their associated costs.

- An urban air quality model, linking pollutant emissions to air concentrations for each of the major air pollutants.

These have to be established for each urban area or air basin. The basic approach in urban air quality management is first to establish from the monitoring network and the air quality guidelines whether current air quality is satisfactory or not during the majority of pollution years. If exceedances of the air quality guidelines are observed, and these are likely to be a general feature each year, then action is required to reduce emissions.

The combination of an air quality model and an emission inventory provides the workhorse of the air quality management process. The aim of the combination is to be able to predict the spatial distributions of the major air pollutants across the air basin. These predictions for each air pollutant can then be compared with the observations from the air quality monitoring network to see whether the major emission source categories have been inventoried satisfactorily and whether the majority of the observed emissions have been accounted for. If the model results are within reasonable bounds of the observations, say within a factor of two, then it should be possible to say whether the current locations of the monitoring sites are giving a reasonable picture or not of air quality across the air basin. The model may indicate the presence of 'hot spots' or grossly polluted locations away from the available monitoring sites. Such locations could be targeted should additional monitoring resources become available or if existing monitoring equipment could be relocated. Mobile monitoring equipment or short-term campaigns are ideal for the identification of 'hot spots' and the rapid confirmation of model predictions.

Policy makers must develop a stream of potential pollution abatement or control measures that can be implemented in the emission inventories of each relevant air pollutant and fed back into the urban air quality model as sensitivity studies. Each of the measures can then be evaluated for its impact on urban air quality and the exceedance of the health protection-based air quality guidelines. The impacts of the individual measures may well vary in magnitude, with some having little impact on air quality and others having a more dramatic influence. An important consideration will be how these sensitivities compare with the general level of model performance for that air pollutant. If the model generally performs well, that is with low bias compared with observations, then we can have more confidence that a particular policy measure will indeed produce the results in the real world that are indicated in the air quality model.

For the urban air quality management process to work, resources need to be given to urban air quality monitoring, emission inventory development and air quality modelling, in a coupled, integrated and consistent manner. Each element is essential and closely interacts with the others. Modelling without monitoring is likely to be meaningless. Emission inventories are an essential element of any urban air quality model, and monitoring can provide an authoritative check on the magnitude and spatial distribution of emission inventories.

6.1.2 Urban air quality modelling

The main function of an urban air quality model is to convert an emission inventory into a time series of air quality concentrations at a number of locations. The emission inventory employed may describe current emissions or reflect some future or hypothetical situation in which emission controls have been implemented or different assumptions made with regard to traffic, industry or population growth or in which certain industries and activities have been completely curtailed.

Urban air quality models aim to describe accurately, and without bias, how pollutant emissions from each source and sector within the inventory influence the spatial distribution of the ground level concentrations of that pollutant, downwind of each source. Emissions from sources close to ground level, such as traffic, exert a much greater influence on ground level concentrations compared with emissions from elevated sources such as chimney stacks. Although the emission inventory quantifies accurately the magnitudes of the emissions

from each emission source and category, their dispersion or mixing characteristics are sufficiently different to make total emissions a wholly inadequate indicator of human health impacts. Whereas simply adding up emissions is meaningless, adding up and integrating over all the ground level concentrations downwind from all the contributing emission sources and categories is precisely what is done in an urban air quality model.

The urban air quality model therefore needs to be able to describe accurately how pollutants disperse away from each source, but it also needs to be able to define which emission sources are upwind and downwind of each location. To do this, all of the information about the individual pollution sources in the emission inventory needs to be accurately spatially referenced. This can be achieved using coordinate systems, such as latitude and longitude, or easting and northings. Monitoring sites and the locations for the output variables all need to be spatially referenced in the same system. However, the concepts of upwind and downwind require a comprehensive understanding of the air flow over the urban air basin and its minute-by-minute variability. Urban air quality models therefore need as a minimum some understanding of how pollutants are transported away from their sources, how pollution plumes are dispersed and how pollutants are brought down to ground level. The track taken through the atmosphere by the pollutants emitted by a pollution source is called a trajectory, in this case, a forward trajectory. The pollutants arriving at a monitoring site may be traced backwards in time to indicate their potential source using back-track trajectories. Urban air quality models need access to meteorological data and compiling them is usually a major undertaking.

As described in Chapter 1, pollutants may be primary (i.e. directly emitted) or secondary. Suspended particulate matter is a mixture of components, some of which are primary pollutants, such as elemental carbon, and some of which are secondary pollutants, such as particulate sulphate and nitrate. Urban air quality models need to represent the chemical processes by which primary pollutants are converted into secondary pollutants as polluted air passes over the air basin.

Urban air quality models are assembled from some or all of the following components:

- A spatially-referenced emission inventory for all relevant primary emitted species and secondary pollutant precursors.
- A dispersion module to describe how ground level concentrations vary down-wind of each source in the inventory.
- An advection module to describe the three-dimensional plume and pollutant trajectories.
- A chemical module to describe how the pollutants react with each other and are transformed into secondary pollutants.
- A deposition module to describe how the pollutants are scavenged from the atmosphere by rain or by removal at the earth's surface or by elements of the vegetation and built environment embedded in the earth's surface.

Because of the large numbers of combinations of individual components that can be potentially assembled to form urban air quality models, it is a difficult task to provide a comprehensive review here. The main types of urban air quality models are described below. The roles played by these models in urban air quality management have also been authoritatively reviewed by Moussiopoulos *et al.* (2003), to which the reader is referred for more details.

6.1.3 Types of urban air quality models

There are two major categories of urban air quality models: empirically or physically based. Empirically-based (or statistical) models are based largely on air quality monitoring data. With the dramatic increase in the number of automatic, continuously recording monitoring sites that have begun operating since the 1990s, empirical models have become more robust and their coverage of different urban environments and air basins has become more comprehensive. These models are used extensively for policy support and for air quality forecasting. Their main strengths are their reliance on underpinning of air quality data.

Their main weaknesses are their assumptions that correlations represent cause and effect and that regression parameters will remain constant into the future.

6.1.3.1 Empirically-based statistical models

To illustrate the approach to urban air quality modelling using empirically-based statistical models, let us consider the air quality problems associated with nitrogen dioxide (NO_2) in large cities. Nitrogen dioxide is a complex urban pollutant because it is both a primary and a secondary pollutant. It is directly emitted into the urban atmosphere by traffic as NO_2 and is formed in the atmosphere by the reaction of primary-emitted nitric oxide (NO) with ozone (O_3). A careful consideration of the results from an air quality monitoring network leads to the conclusion that there is an empirical relationship between the annual mean NO_2 concentrations and the annual mean $NO_x = NO + NO_2$ concentrations measured simultaneously (Stedman *et al.*, 1997), of the form:

$$[NO_2] = a \, [NO_x]^b$$

This relationship has been constructed from air quality data for a range of site locations from heavily polluted road sides, to suburban or rural locations far removed from traffic. The empirically-based model comprises the constants a and b and would be most easily handled using a spreadsheet.

All things being equal, this relationship should describe how NO_2 concentrations may decline in the future if NO_x levels were to decline in response to pollution controls. Because the exponent is less than unity, NO_2 levels should decline less rapidly compared with NO_x levels and this non-linearity in the response of NO_2 to NO_x emission reductions would be a direct consequence of the primary and secondary behaviour of NO_2. In this way, policy makers can be advised how much NO_x emissions would have to be reduced to produce acceptable air quality of NO_2 in the future (Stedman *et al.*, 1997). Embedded in the above empirical relationship is the current availability of ozone from the rural background surrounding the urban area. If this is increasing, perhaps due to the hemispheric scale increase in ozone, then all things will not remain equal and the parameters a and b may be expected to change in the future (see also Chapters 3 and 4).

For a further example of the approach to urban air quality modelling using empirically-based statistical models, we turn to the problem of issuing an air quality forecast for an air basin. An empirically-based statistical approach to air quality forecasting would entail fitting an observed time series of pollutant concentrations for a period in the past to a set of parameter values covering the same time period, using some statistical procedure. The results of the fitting procedure in terms of coefficients scaling each of the input parameters would then comprise the urban air quality model. The forecast would be generated by compiling a synthetic set of parameter values for the future situation and feeding them back into the statistical model and thereby estimating the pollutant concentration. The selection of the parameters to be used in the generation of the statistical model is the most difficult issue. Clearly, they should include some consideration of meteorological and dispersion parameters, such as wind speeds and directions, boundary layer depths, temperatures and humidities. The selected parameters also need to take into account some aspects related to emissions, such as traffic flows, time of day, seasons and (again) temperatures because these may influence space heating and cooling demands.

Empirically-based statistical urban air quality models have a number of important advantages. They require only simple computer support, being mostly spreadsheet-based once they have been built. They obviate the need for complex descriptions of the emissions, transport and transformations of air pollutants and focus almost entirely on air quality data for their underpinning. They are straightforward to verify and generally command a great deal of confidence from policy makers as a consequence. On their own, however, they are difficult to defend scientifically without the support of other modelling activities that can convincingly link the selected causal parameters to the modelled effects.

6.1.3.2 Physically-based (deterministic) models

Physically-based deterministic urban air quality models vary enormously in sophistication, type and application and include Gaussian plume models, box models, Eulerian grid models, Lagrangian trajectory models and computational fluid dynamics (CFD) models. They are all based on some form of the mathematical representation of the underlying processes and rely on the availability of observations to provide verification, evaluation and validation. They are generally highly computer-intensive and require large amounts of input data to run effectively.

Gaussian plume models Gaussian plume models are based on simple solutions of the diffusion equation and are widely used to describe the dispersion and advection of primary pollutants away from single or multiple pollution sources. Using fixed plume geometries, they offer a practical and physically-based treatment of the dispersion of pollutants emitted from large point sources, such as power stations; area-based sources, such as commercial, industrial and domestic sources and from linear sources, such as motor vehicle traffic. It is generally assumed that the ground level concentration at any point may be obtained by simply summing the contributions from each of the individual sources in the emission inventory to the ground level concentration at that point. The models can represent many of the important chemical processes and can treat most of the important spatial and temporal scales in the urban environment. Furthermore, their output is in the form of maps of air quality levels and air quality guideline exceedances so they provide a ready visualization for policy makers of the impacts of policy measures and controls. They are used extensively for policy applications and regulation on the local scale. Gaussian plume models have been applied to the calculation of long-term average concentrations for a city the size of London (Chamberlain *et al.*, 1979). Examples of these models are readily obtained from the US EPA website http://www.epa.gov/scram001/tt22.htm.

Screening models are simplifications of Gaussian plume models and are designed to provide an initial estimate of pollutant concentrations arising from specific sources, particularly traffic. They require relatively simple input and provide an assessment of whether a more complete analysis is required or not. Although they are designed to be quick and simple to use, it should not be assumed that they are necessarily excessively inaccurate, but they do take a necessarily cautionary approach. Screening models have their main application on the micro-scale in the assessment of 'hot spots' within urban road networks.

Box models Box models are a further variant or simplification of the Gaussian plume approach. The basic formulation of a box model assumes that clean air is carried across a city of dimension, d, as pollutants are emitted into it. From a mass balance between the pollutants emitted and those carried away by the wind through the downwind vertical face of the box, then:

$$c = Q / d h u$$

where Q is the emission rate in g s^{-1}, h is the depth of the box and u is the wind speed (Derwent *et al.*, 1995). The emission term Q requires the treatment of traffic, space heating and other emission sources with some allowance for their diurnal and seasonal patterns. The depth of the box, and the wind speed determine how the urban air concentrations respond to the meteorological parameters. Poor vertical mixing (small h) and light winds (small u) indicate poor air quality (large c) when emissions are high (large Q).

Screening and box models are not highly computer-intensive and can be implemented as simple spreadsheet tools.

Eulerian grid models Eulerian grid models are the most commonly used approach to urban air quality modelling. In this approach, a regular three-dimensional grid mesh is assembled throughout the urban air basin and all processes are presented at the intersection points of the horizontal and vertical meshes. These models generally include a full meteorological model to calculate the mean flow and the turbulence for the transport and advection of the air pollutants and to describe the dispersion of pollutants within the flow. They gener-

ally require large amounts of input data and computer time. They also produce large amounts of output making them difficult to interpret and to understand. Some applications to western European cities are described in Section 6.1.5.2.

Current computer capacities necessarily limit the smallest horizontal grid sizes to about 1 km × 1 km and the time resolution to hourly, so that integration periods are generally limited to a few days or weeks and it is not always practical to calculate long-term averages. The great advantage of these models is their ability to handle time-varying meteorology in a detailed manner, making them ideal for the analysis of pollution episodes (see Chapter 5). Furthermore, they can include secondary pollutants as well as primary pollutants. Thus, they can easily handle a wide range of air quality problems with the same model addressing urban air quality, photochemical smog formation, acid rain, eutrophication and the build-up of persistent organic pollutants in the aqueous and terrestrial environments.

The output from these Eulerian models is available for each model time-step as concentrations at the intersections of the grid meshes. These can be readily mapped and presented in movie format to provide a vivid representation of the impact of control strategies for policy makers. Because of their widespread application now throughout the world, Eulerian grid models represent the state-of-the-art in urban air quality modelling. A review of these Eulerian grid-based urban air quality models is given in Moussiopoulos *et al.* (1996) which is available on the European Environment Agency website http://www.reports.eea.eu.int/92-9167-028-6/en/tab_content_RLR.

Lagrangian trajectory models Lagrangian trajectory models adopt an altogether different coordinate framework compared with Eulerian grid models, by representing the air flow across an urban basin by the advection of air parcels in a three-dimensional wind flow. All the processes are represented on the moving air parcels rather than relative to a fixed Eulerian grid. Lagrangian models offer the prospect of much more realistic treatments of turbulence, convection, advection and chemical processes influencing urban air quality. They share many of the advantages and disadvantages of complex deterministic models. They require large amounts of input data and the results of their verification and validation studies are difficult to interpret. Their single great advantage is the economy and efficiency of the Lagrangian advection scheme over its Eulerian counterpart which allows significantly greater chemical sophistication to be used in the chemical mechanisms employed in Lagrangian models.

Computational fluid dynamics (CFD) models Computational fluid dynamics (CFD) models also solve the fluid dynamic equations implicit in the detailed description of the transport of air pollutants over the urban environment. Examples are CFX, FLUENT and STAR-CD (see Galpin *et al.*, 1985). Their focus is on the detailed modelling of the complex structures at the micro-scale generated by the air flow over buildings. These include the vortices and recirculations found in street canyons and the dead spaces found around buildings which have a huge influence on urban air quality and the occurrence of 'hot spots'. They are thus able to aid in the understanding of local micro-scale processes but are not appropriate for calculating concentrations of pollutants across urban air basins.

6.1.4 Examples of urban air quality models

It is not practical to try to provide a comprehensive list of urban air quality models here, but Table 6.1 provides an introduction to 29 urban air quality models that have been implemented in Europe over the last ten years. The summary is taken from the review of Moussiopoulos *et al.* (1996).

Many of the models in Table 6.1 are classed as Gaussian plume models: AUSTAL 86, CAR-FMI, CONTILINK, CTDM PLUS, HPDM, IFDM, INPUFF, ISCST-2, OML, PLUIMPLUS, ROADAIR, SCALTURB and UK-ADMS. The majority of these applications involve the study of a single primary pollutant being emitted from a single chimney stack and its dispersion

in the downwind environment. This is an important application for the regulation of atmospheric discharges of harmful, toxic and radioactive gases and particles. These models are generally considered to be reliable over downwind distances of 10-20 km and this reliability is improved if local meteorological data are available. Model verification and validation is usually achieved through the analysis of field trials using tracer releases.

Traffic makes an important contribution to urban air quality in most towns and cities and it has been the subject of a large number of modelling studies. Table 6.1 lists the following traffic models: CAR, CAR-FMI, CONTILINK, OSPM and ROADAIR. These models are often combinations of Gaussian plume and empirical models and are frequently used for screening purposes. They are used in the design of road layouts and junctions and in the zoning of housing to minimize exposure to motor vehicle pollution. These models are usually simple to use and are considered reliable over distances of within 100–250 metres from the road system. Model verification and validation usually relies on the establishment of instrumented road side sites and the observation of the three-dimensional distribution of pollutants in the vicinity of heavily-trafficked roads.

It has proved difficult to build atmospheric chemistry into the Gaussian plume approach and so the approach favoured for the study of secondary pollutants, such as ozone, has been the Eulerian grid model (see Table 6.1). The CALGRID, CIT, DRAIS, MARS and UAM are all examples of three-dimensional grid models in which the processes of atmospheric dispersion, transport and chemical transformation are represented on a regular grid mesh, with horizontal dimensions of 1 km or so.

Eulerian grid models are often composed of a meteorological model and a number of other modules describing emissions, atmospheric chemistry and deposition processes. They may be assembled within the framework of the meteorological model, online, or run separately from the meteorological model, in which case they are called off-line models. They are generally highly computer-intensive and can be difficult to verify and validate without dedicated field campaigns and experimental programmes. Table 6.1 lists a number of meteorological models that have been used in urban air quality modelling either as a central part or as an off-line model used to drive an Eulerian grid model. These include: MEMO, RAMS and TVM.

The models DISPERSION, EPISODE, HYPACT, MATCH and UDM-FMI have been

TABLE 6.1 Examples of urban air quality models based on the review of Moussiopoulos *et al.* (1996)

Name	Type of model	Application
AUSTAL 86	Gaussian plume	Primary pollutants
CALGRID	Eulerian grid	Secondary pollutants
CAR	Empirical	Traffic pollutants
CAR-FMI	Gaussian plume	Traffic pollutants
CIT	Eulerian grid	Secondary pollutants
CONTILINK	Gaussian plume	Traffic pollutants
CTDM PLUS	Gaussian plume	Primary pollutants
DISPERSION	Empirical	Urban air quality
DRAIS	Eulerian grid	Secondary pollutants
EPISODE	Eulerian grid	Urban air quality
HPDM	Gaussian plume	Primary pollutants
HYPACT	Eulerian grid	Urban air quality
IFDM	Gaussian plume	Primary pollutants
INPUFF	Gaussian plume	Primary pollutants
ISCST-2	Gaussian plume	Primary pollutants
MARS	Eulerian grid	Secondary pollutants
MATCH	Eulerian grid	Urban air quality
MEMO	Eulerian grid	Meteorology
MERCURE	Eulerian grid	Meteorology
OML	Gaussian plume	Primary pollutants
OSPM	Empirical	Traffic pollutants
PLUIMPLUS	Gaussian plume	Primary pollutants
RAMS	Eulerian grid	Meteorology
ROADAIR	Gaussian plume	Traffic pollutants
SCALTURB	Gaussian plume	Primary pollutants
TVM	Eulerian grid	Meteorology
UAM	Eulerian grid	Secondary pollutants
UDM-FMI	Eulerian grid	Urban air quality
UK-ADMS	Gaussian plume	Primary pollutants

used intensively for urban air quality management in Europe. These applications usually require the treatment of multiple sources of pollutants and spatial scales of up to 25-50 km.

6.1.5 Urban air quality management in action

6.1.5.1 Santiago de Chile

Where data are relatively sparse, it can be appropriate to apply box models to urban air quality; this has been done in the case of the city of Santiago in Chile (Jorquera, 2002a and b). Santiago lies in a topographic basin between the Coastal and Andean mountain ranges. It has a dry climate and thermal inversions are frequent, especially in the autumn and winter (GEMS/AIR, 1996). The city has a population of nearly 6 million (about 40 per cent of the national total), houses some 70 per cent of the country's industry and has about 800,000 cars and buses (Romero *et al.*, 1999). High insolation, poor dispersion and substantial emissions make the city prone to poor air quality. There are also problems with resuspended dust from dry soils. In 1992, Santiago was declared 'saturated' with respect to SO_2 and in 1996 the same situation was identified for PM_{10}, ozone (O_3), suspended particulate matter and carbon monoxide (CO). A clean-up plan was launched in 1998 (revised 2001) and there has been an overall improvement in air quality. There are no spatially disaggregated emissions inventories for the city, but there are totals by sector. These data are available via the Internet (http://www.conama.cl/rm – in Spanish) and are summarized in Table 6.2.

The city does have quite a good air quality automatic monitoring network which has been in place since 1988 under the auspices of SESMA (Servicio de Salud del Ambiente Region Metropolitana – Environmental Protection Service for the Metropolitan Region), and a network of 22 meteorological monitoring stations.

Using the data available, box models have been developed for Santiago to assess the contribution of different source sectors to the overall pollution load (Jorquera, 2000a and b). The models are based on least squares regression of monthly measured pollutant concentrations against estimates of monthly emissions from the major sources using emission factors. For CO, emissions were assumed to be proportional to sales of leaded and unleaded petrol, for NO_x to sales of leaded petrol, diesel and fuel oil and for SO_2 fuel oil and kerosene (higher S content than other fuel types). Regressions are carried out for each monitoring site in turn to estimate the contribution from the different source types to pollution at those sites. The models include a dimensionless dispersion factor which can account for seasonally varying dispersion conditions. The modelling exercise has shown up the effects of changes in the overall age of the vehicle fleet (newer vehicles are less polluting than older ones), that local sources (especially diesel vehicles) dominate NO_x concentrations in the city centre and that SO_2 is rather different, with stationary sources playing a larger role. The models can also be used to estimate seasonal patterns of emissions reductions needed to improve air quality in the winter. In the case of CO and NO_x, emissions from vehicles in the winter would need to be

TABLE 6.2 Emissions (tonnes/yr) for Santiago in 2000 (Source: Comisión Nacional del Medio Ambiente (CONAMA)

Source	PM_{10}	CO	NO_x	VOC	SO_x	NH_3
Stationary	2.6	16.0	8.1	56.3	6.0	28.4
Mobile	2.4	175.6	46.6	24.7	3.1	0.9
Total	5.0	191.6	54.7	81.0	9.1	29.3

reduced by some 75 per cent in order to achieve the same concentrations as occur in the summer.

Modelling of particulate matter (a major problem in Santiago) has followed the same general approach, estimating primary PM_{10}, $PM_{2.5}$ and coarse particle emissions as fractions of CO and SO_2 emissions, but including a term for emissions from forest fires, construction etc. Secondary aerosol production is not included. The models indicate that vehicles are the main source of particulates in Santiago and have become increasingly important in their contribution to the fine fraction ($PM_{2.5}$). This type of modelling approach is clearly valuable in urban areas where the range of input data required by more complex models is not available. It was apparent from the Santiago exercise, however, that the model parameters may not be applicable over long time periods as factors such as changes in fuel composition, age of vehicle fleet and adoption of technologies (such as catalytic convertors) come into play.

6.1.5.2 Cities in western Europe

Across Europe, many cities have the range of emissions, meteorological, monitoring, topographic and land use data needed to underpin complex models. Such models attempt to include as many relevant physical and chemical processes as possible and usually take the form of Eulerian 3-D models. As well as having large data requirements, such models are computationally-intensive and can sometimes only be run for short periods of time (e.g. a few days, rather than a full year). It can, therefore, be desirable to have simpler models where more processes are parameterized, or use empirical relationships. These simpler models can be run over longer time periods and more often. As part of the EU's Auto-Oil Programme, air quality has been modelled for a number of European cities using models of varying complexity. As part of the main Auto-Oil Programme II study (AOP-II), three complex air pollution models were applied to ten cities from Dublin and Madrid in the west to Helsinki and Athens in the east (CEC, 2000). In a second study, simpler models were applied to about 200 urban centres across the EU to

estimate background urban air quality for a reference year (1990 or 1995) and in 2010 (EEA, 2001; de Leeuw, 2001). The outputs were then used to estimate the proportion of the EU population exposed to potentially dangerous levels of pollution.

The AOP-II project incorporated modelling at regional, urban and street canyon levels using four spatial scales (30×30 km, 6×6 km, 2×2 km, canyons). Urban canyon modelling was only applied to Berlin and Milan. Three dispersion models were used: CALGRID, Urban Airshed Model (UAM) version IV and the European version of the Urban Airshed Model (EUAM), all of which are 3-dimensional photochemical models (see Table 6.1). EUAM includes an explicit treatment of plume dispersion, a process not handled well in many Eulerian models. Hourly meteorological data were provided by three different models initialized using measurements (surface and upper air) and information from the European Centre for Medium Range Weather forecasting. The pollutants of interest were CO, NO_2/NOx, benzene, O_3 and PM_{10}. Given the complexity of the models, it was not feasible to run them for a full year, so shorter periods were chosen to represent long averaging periods to yield surrogates for annual means and estimates of 98th percentiles of 1-hr means.

6.2 Regional air quality

6.2.1 Regional air quality management and policy

Concerns about regional air pollution issues are much more recent than urban air pollution issues and have been raised from the 1970s onwards. Initially, these concerns focused on acid rain and the long-range transboundary transport of acidifying sulphur compounds to the remote, sensitive ecosystems in Scandinavia (see Chapter 4). The concerns were such that the Geneva convention was agreed, under the aegis of the United Nations Economic Commission for Europe (see Chapter 7), to control and reduce the transboundary transport of acidifying agents from the countries of north-west and central Europe to Scandinavia (Wuster, 1992).

At the heart of this regional air pollution policy development is a regional air quality modelling system which has been employed to quantify the export of acidifying sulphur compounds from the emitting countries and their import into the receptor countries of Scandinavia where the impacts of acidification were felt to be at their greatest. This regional air quality modelling system is the European Monitoring and Evaluation Programme (EMEP) model (Eliassen and Saltbones, 1983).

During the 1980s, it was realized that sulphur compounds were not alone in causing acidification and that both oxidised and reduced nitrogen compounds derived from NO_x and ammonia emissions were also contributing. Nitrogen oxides and ammonia were then added to the EMEP regional air quality model so that it could be used to quantify the exports and imports of acidifying and eutrophying nitrogen compounds from each European country. Then, because of concerns about forest dieback and human health effects, the attention of policy makers shifted to ground level ozone and its regional scale formation and transboundary transport. To support regional scale ozone policy formulation and development, the EMEP model was extended to deal with ozone precursor emissions and the transboundary transport of ozone in regional pollution episodes.

More recently, during the late 1990s, the concerns of European policy makers have shifted to address the formation and transport of fine particulates on the regional scale and their impact on human health. Fine particulates have primary components such as elemental carbon, emitted by combustion sources, and secondary components such as particulate sulphate and nitrate, formed from the oxidation of SO_2 and NO_x. Because of their long lifetimes, fine particulates undergo transboundary transport and so their effective control also requires concerted action on the regional scale.

Policy actions to control acidification, eutrophication, ground level ozone and fine particles interact so strongly because of chemical interactions on the regional scale. As a consequence, policy makers have agreed to approach European regional air pollution problems in a holistic and integrated manner. Fig. 6.1 shows how the various air pollution problems are interrelated: because they share common sources of pollution; or because of chemical interactions; or because they produce similar environmental effects; or because they share common receptors. The multi-pollutant and multi-effect approach to regional pollution problems has been pioneered within Europe and has been made possible by the availability of regional air quality models that can address the interactions between the pollutants indicated in Fig. 6.1.

6.2.2 Regional air quality modelling

To enable regional air quality models to describe the interactions inherent in Fig. 6.1, all models, whether Lagrangian trajectory models or Eulerian grid models, share common elements, modules and databases, as follows:

- Emissions data, by country and by grid square of SO_2, NO_x, NH_3, organic compounds, primary particles, CO and methane from the indicated anthropogenic sources and from natural biogenic sources.
- Meteorological data, comprising wind speed and directions, boundary layer depths, boundary layer turbulence, precipitation, cloud properties, solar radiation.
- Deposition module, describing the scavenging of pollutants by rain and the uptake of pollutants by the underlying surface.
- Chemistry mechanisms, capable of describing the photochemical formation of ozone and fine particles and the formation of acidifying and eutrophying agents in cloud droplets.

The main sources of information on emissions are national activities. These sources are brought together on a European basis through the activities of the European Topic Centre on Air Quality of the European Environment Agency (on behalf of the European Union) and the EMEP Task Force on Emission Inventories

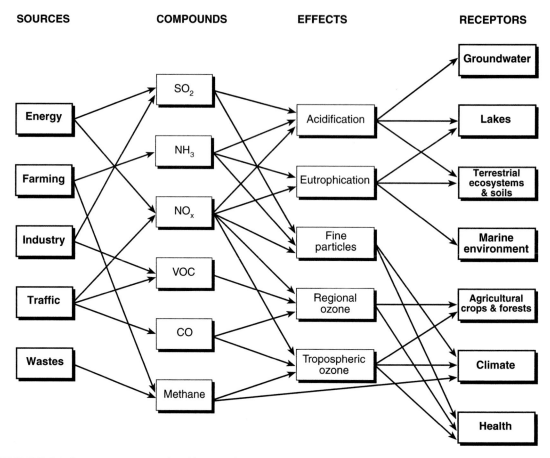

FIGURE 6.1 Sources, compounds, effects and receptors for each of the regional air pollution problems being tackled in Europe during the 1990s and 2000s

and Projections (on behalf of the United Nations Economic Commission for Europe) (see Chapters 2 and 4). On a global basis, national inventories feed through to the Intergovernmental Panel on Climate Change and to independent published inventories compiled elsewhere.

Almost all of the meteorological data required by regional air quality models are provided by meteorological models of one form or another. Numerical weather forecast models are an important source of input data for regional modelling, where the regional model data are a by-product of the weather forecast production system. Dedicated meteorological models are also an important source of input data required by regional models. Here the

dedicated meteorological model acts as a driver for the regional air quality model and both models will often share a great deal of infrastructure between them, such as model domains, coordinate systems, map projections and grid meshes. Dedicated meteorological models will need access to numerical weather forecast models for their initialization but will be 'free-running' once they have started.

Most regional air quality models use much the same chemistry mechanisms and deposition modules and there is generally little to choose between the model approaches and between those adopted within urban air quality models. The main issue here is not the choice of parameters within these modules, but how these modules are driven by the regional air quality model. For example, with the representation of cloud chemistry in the different models, the main issue is the accuracy of the cloud amount and cloud liquid water content provided by the meteorological model rather than with the chemical kinetic parameters themselves. The cloud properties provided by the meteorological model may be quality assured for utilization within the meteorological model, but may not be accurate and reliable enough to drive the cloud chemistry module in an altogether different regional air quality model. Boundary layer depth is another output from the meteorological models which may not be accurate and reliable enough to drive the deposition modules in regional air quality models under all atmospheric conditions.

The main differences between Eulerian urban and regional air quality models is the important focus in the former on atmospheric dispersion and diffusion and in the latter on chemical interactions between the different pollutants. On the urban scale, the interest is primarily on how pollutant emissions are advected away from sources and on how turbulent transport brings pollutants down to ground level. By the time the polluted air masses leave the urban area, the dispersion processes have efficiently mixed the pollutants in the vertical and a well-defined urban plume has been formed in the horizontal. For this reason, detailed understanding of turbulent transport is crucial on the urban scale but less so on the regional scale. However, with the longer travel times on the regional scale, there is much greater time for chemical interactions to take place and hence there may be substantial chemical development within regional pollution plumes. Consequently, chemical interactions tend to be more important on the regional scale compared with the urban scale.

6.2.3 Examples of regional air quality models for Europe

Initially, the main regional air quality model within Europe was the EMEP model, as outlined in the historical development above. This situation has changed more recently with the development of a number of Eulerian grid-based regional air quality models. Table 6.3 provides an introduction to 12 regional air quality models that have been implemented in Europe over the last ten years.

In the main, the regional air quality models in Table 6.3 are verified and validated by comparison of model results with observations from national networks or from the EMEP network. For key pollutants, such as acidic species, ozone and its precursors, the EMEP network (see Chapter 2) provides daily or hourly observations. National networks provide higher spatial resolution and cover the more polluted rural, suburban and urban locations. So a combination of EMEP and national monitoring data are often used for model testing and evaluation. Intensive field campaigns provide observations of a far wider range of pollutants and trace gases for model evaluation, but cover smaller spatial regions and for limited periods.

The Chernobyl nuclear accident in 1986 provided an opportunity for the testing and an evaluation of a number of regional scale long-range transport models within the Atmospheric Transport Model Evaluation Study (ATMES) experiment (Klug et al., 1990). The European Tracer Experiment (ETEX) evaluated the performance of a number of long-range transport models using observations following the atmospheric release of inert tracers.

The main purpose of the models in Table 6.3 is to evaluate policies for the control of acidic deposition or of ground level ozone. Because acidic loads and ozone exposure levels need to be assessed over extended time periods, these models are often run for periods of up to ten years or so. There are therefore practical limits to the extent of the details that can be built in to the representation of the basic chemical and physical processes in them. Furthermore, these models are required to operate on the European scale and this places lower limits on the spatial scale required for the representation of the basic transport and emission processes to about 50 km × 50 km. Consequently, it is difficult to represent processes such as the orographic enhancement of precipitation, coastal sea breezes, mountain valley flows and urban heat island effects.

TABLE 6.3 Examples of the application of regional air quality models in Europe based on the review of Moussiopoulos *et al.* (1996)

Model	Type of model	Application
DMU	Eulerian grid	Acid rain, ozone
EMEP MSC-E	Eulerian grid	Heavy metals, persistent organic pollutants
EMEP MSC-W	Eulerian grid	Acid rain, ozone
EURAD	Eulerian grid	Acid rain, ozone
GKSS	Lagrangian trajectory	Heavy metals
HARM	Lagrangian trajectory	Acid rain
IVL	Lagrangian trajectory	Ozone
LOTOS	Eulerian grid	Ozone
REM3	Eulerian grid	Ozone
TREND	Statistical	Acid rain
UiB	Eulerian grid	Ozone
UK PTM	Lagrangian trajectory	Ozone

6.2.4 Regional air quality modelling in Europe

The models developed through EMEP have played a key role in developing emissions reduction policies for Europe within the context of the United Nations Economic Commission for Europe (UNECE) and the European Union (EU) (see Chapter 7). Part of the EMEP remit is to provide estimates of concentrations, depositions, transboundary fluxes and budget matrices. As well as synthesizing measurement data, EMEP has developed and implemented 2-dimensional Lagrangian and 3-dimensional Eulerian models for acidification, eutrophication, ground level ozone and, most recently, heavy metals. The EMEP Lagrangian Acid Deposition Model (LADM) was used from 1985 for sulphur only, but from 1987 oxidised and reduced nitrogen were included in the coupled chemical scheme. Some early applications are described in Eliassen and Saltbones (1983). The LADM is a receptor orientated trajectory model with a spatial resolution of 150 km × 150 km and one vertical layer the depth of the mixing layer. It has a coupled chemical scheme for ten compounds. Meteorological inputs are taken

for every six hours from one of the Norwegian Meteorological Institute's Numerical Weather Prediction models, currently LAM50E. LADM is described in detail by Tsyro (1998) and at http://www.emep.int/acid/ladm.html. Model performance was assessed by comparison with measurements from the EMEP monitoring network. As well as serving as a form of model validation, this also allowed the impacts of emissions reductions since 1985 to be assessed.

Given the significant role played by LADM in policy development, its replacement by an Eulerian model was carried out gradually. The effects of this transition have been described by Tarrason *et al.* (2001). Examples of the output from the EMEP Eulerian model for 1999 are shown in Fig. 6.2. Since 2003, a unified modelling system has brought together acidification and oxident modelling (Simpson *et al.*, 2003). Further details are available from the EMEP website http://www.emep.int.

Modelled depositions are used in combination with critical loads maps to identify areas of likely ecosystem damage whether through excess inputs of acidifying pollutants or excess nutrients (see Chapter 4). Source-receptor matrices allow the contributions of each EMEP

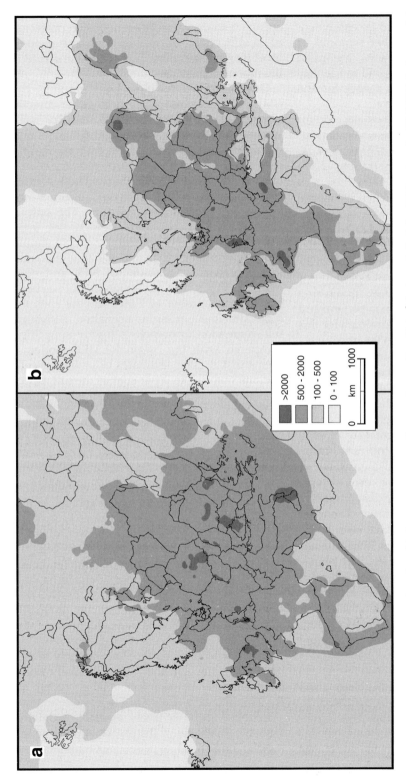

FIGURE 6.2 Sulphur (a) and reduced N (b) deposition across Europe (in mg S or N/m^2/yr) modelled for 1999 using the EMEP Eulerian model in units of g N or S m^{-2} yr^{-1} (with permission from the Norwegian Meteorological Institute/EMEP/MSC-W)

country to deposition within their own boundaries and also other countries to be estimated. This helps to determine whether national or international emissions reductions are most likely to reduce pollution levels and to identify which countries might be the target for the most stringent emissions controls.

An EMEP Lagrangian photochemical oxidant model was developed by Simpson (1992, 1993) and is described in detail in Simpson *et al.* (1997). Modelled ozone concentrations have been compared with measured values for both daily maxima and for 12 noon. When values are compared for 12 noon GMT averaged over April to September, there is a good relationship between them, both data sets showing a similar gradient across Europe. The largest model overpredictions occur over north-west Spain, but it should be noted that there are few monitoring points around the Mediterranean in general and it is not clear whether the emissions data for this region are of the same quality as for north-west Europe. Data and model are also compared for the two recognized indicators of ozone damage: AOT_{40} (crops) and AOT_{60} (health).

Although Lagrangian models formed the basis of the EMEP modelling effort for many years, the simplified vertical structure of such models was recognized as a significant limitation, particularly for the representation of pollutant transport above the atmospheric boundary layer and the vertical distribution of pollution within the boundary layer. The most recent activities within EMEP have been in the development of a unified model which brings together the Eulerian acidification and photo-oxidant models. The aim has been to put together a core set of subroutines and modules which can then run optional chemical schemes. This type of integrated and modular modelling approach is already in use (e.g. by the United States Environmental Protection Agency) and is discussed further below in Section 6.2.5.

The EMEP models have not been used directly for scenario analysis, but their source-receptor matrices are used within the framework of IIASA RAINS. RAINS (Regional Air Pollution Information System) is an integrated assessment model which takes a modular approach to the regional air pollution problems of acidification, eutrophication and ground level ozone formation. The overall structure is illustrated in Fig. 6.3.

One of these modules, DEP, employs the EMEP source-receptor matrices either directly (for acidification and eutrophication) or indirectly (for ozone). For acidification and eutrophication (S and N deposition) RAINS uses long-term average source-receptor matrices taken from the EMEP model. These source-receptor relationships between countries are assumed to hold into the future. The output from the EMEP ozone models is not used directly, but has been used to develop a simple regression model which relates ozone exposure to the precursor emissions of NO_X and non-methane VOCs. The regression model has been based on a number of EMEP model runs for high and low emissions scenarios. Although this regression approach is highly simplified compared with the original EMEP model, it seems that it gives estimates of ozone concentrations within 1 to 2 per cent of the full model (Heyes *et al.*, 1996). Scenario runs with the full EMEP unified model are planned.

RAINS has become the primary modelling tool for policy development within the UNECE and EU. This is because depositions or concentrations can be set to meet particular environmental (including health) targets expressed by metrics such as critical loads, critical levels, AOT_{40}. Using the transfer matrices to provide source-receptor relationships, the optimization module (OPT) then allocates emissions reductions in a manner that minimizes costs. Reflecting the overall pattern of changing concerns over air pollution and effects, RAINS was first developed for sulphur and then extended to include oxidised and reduced N and later ozone. The most recent change has been to model particulate matter. The impacts of the implementation of the Gothenburg Protocol (see Chapter 7) on mean AOT_{60}, based on RAINS output, is shown in Fig. 6.4

RAINS-Europe is freely available from the IIASA website (http://www.iiasa.ac.at/~rains/index.html.

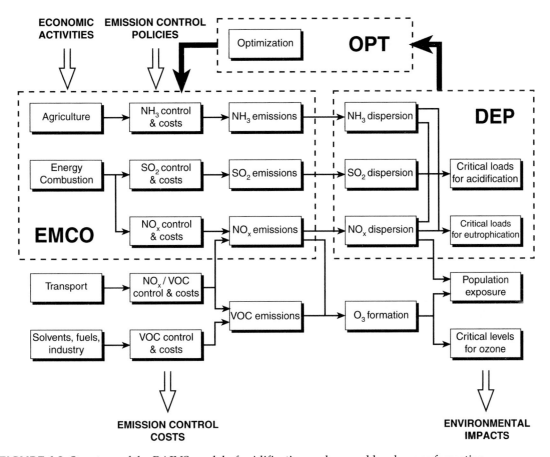

FIGURE 6.3 Structure of the RAINS model of acidification and ground level ozone formation

6.2.5 Regional air quality modelling in North America

Photochemical smog, and ozone in particular, have been a major air quality concern in the USA over the last 50 years or so. Initially, these concerns were limited to the Los Angeles Basin in southern California (see Chapter 5) but now there are many regions of the USA that do not meet air quality standards for ozone, set to protect human health. In the past, individual states have used the United States Environmental Protection Agency (US EPA) Empirical Kinetic Modelling Approach Ozone Isopleth Model (EKMA-OZIPM) to investigate the reductions in ozone precursor, NO_x or VOCs, emissions that would be required for them to meet the ozone air quality standard at

some time in the future. The EKMA-OZIPM model is based on a simple box model that moves over an urban area, picking up precursor emissions, then passes into the downwind areas as ozone concentrations rise due to photochemical formation (Gery and Crouse, 1990). The model predicts how the maximum ozone concentrations reached downwind during one day's photochemistry are related to the initial ozone precursor concentrations found during the early morning in the centre of the urban area. When the maximum ozone concentrations are plotted out against the initial NO_x and VOCs concentrations, a set of isopleths are produced. From this diagram, it is possible to work out for each urban area whether NO_x or VOCs control is the more appropriate ozone control strategy.

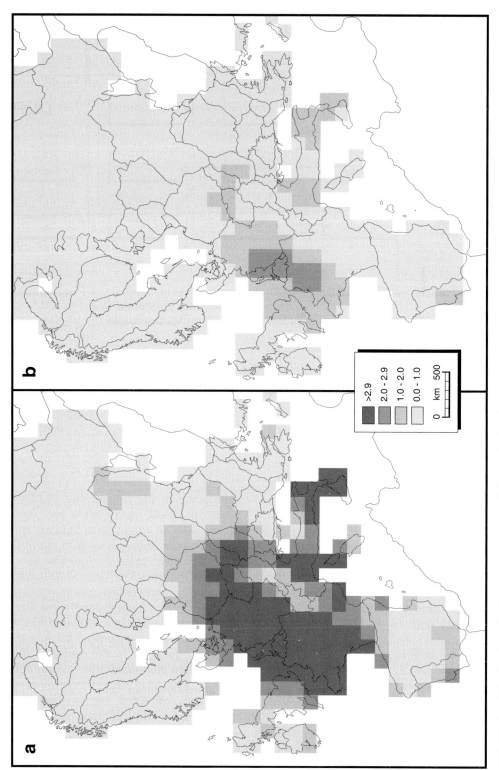

FIGURE 6.4 Distribution of AOT_{60} exposures in ppb hours for (a) the RAINS 1990 scenario (taken from the IIASA website) and (b) the 2010 Gothenburg Protocol scenario

Increasingly, it has been realized that as well as efficient urban scale ozone formation, long-range transport can play an important role in producing the elevated ozone levels that cause the exceedance of air quality standards. Long-range transport is understood to be important in the Los Angeles Basin and along the eastern seaboard of the USA, where the area of ozone non-attainment stretches from Washington DC, northwards to Boston and beyond. The processes responsible for ozone long-range transport are difficult to describe in the simple box model approach of EKMA-OZIPM. A number of complex 3-D Eulerian grid models have been developed to study photochemical ozone formation and its long-range transport. These models are summarized in Table 6.4.

While the modelling effort for the regional scale within Europe is only now moving from a Lagrangian to an Eulerian approach, this has not been the case in North America, as Table 6.4 shows. The Regional Acid Deposition Model (RADM) was developed during the 1980s, for both regional scale ozone and acid deposition modelling (Stockwell *et al.*, 1990). This integrated approach was adopted more than a decade ahead of Europe.

The US EPA has changed its modelling strategy and set up the Community Modeling and Analysis System (CMAS) which is built around the Models-3 Eulerian grid model. The central aim of the CMAS concept is to have US EPA scientists and policy makers, university scientists, regional policy makers and industry, all using the same Models-3 system. Models-3 adopts a modular approach enabling the user to switch between different state-of-the-art physical parameterizations, chemical mechanisms or process descriptions. The CMAS system is web-based and further details are available from: http://www.cmascenter.org.

6.2.6 Regional air quality modelling in Asia

As concerns over regional scale air pollution issues in Asia have emerged since the 1990s (see Chapter 4), modelling efforts have been able to learn from experience gleaned in North America and Europe. In some cases, models developed for one part of the world have been adapted to operate elsewhere.

Carmichael *et al.* (2002) report on an intercomparison exercise for sulphur involving eight long-range transport models (both Lagrangian and Eulerian). The models were all run using the same SO_2 emissions and a common set of meteorological data. The intercomparison had four elements: modelling SO_2 and SO_4 concentrations and sulphur deposition using the models' standard parameter sets; running using specific parameter values for the conversion of SO_2 to SO_4 and for wet removal of SO_2 and SO_4; modelling source-receptor relationships for a given number of sites and allowing the modellers to tune their models once they had seen the measurement data. All the models were able to reproduce the overall patterns of observed concentrations and deposition, although most overpredicted SO_2 concentrations and underpredicted SO_4

TABLE 6.4 Urban and regional ozone air quality models developed in the USA

Model	Type of model	Reference
EKMA-OZIPM	Box model	Gery and Crouse (1990)
CALGRID	Eulerian grid	Yamartino *et al.* (1992)
CIT	Eulerian grid	McRae *et al.* (1982)
Models-3	Eulerian grid	www.cmascenter.org
RADM	Eulerian grid	Chang *et al.* (1987)
UAM	Eulerian grid	Ames *et al.* (1985)

concentrations and wet deposition. There were a few monitoring sites for which all models performed badly and the authors suggest that this might be due to a combination of uncertainties regarding local processes (including emissions) and the horizontal resolution of the models. Having the correct amount of precipitation in the models was found to be crucial. Model performance was also found to improve the longer the period over which data and models were compared. This reflects the difficulty of capturing episodicity. The comparison of source-receptor matrices, so important in the policy context, showed that the models were generally consistent in their identification of the major source area. This exercise did, however, indicate the importance of vertical structure during the winter monsoon when winds at different levels may come from very different directions. Overall, this exercise found that emissions and meteorological factors are the most important determinants of deposition.

RAINS-ASIA has been using S depositions calculated using the ATMOS model. This is a multiple layer Lagrangian model which treats emissions of SO_2 as puffs which then mix horizontally following a Gaussian distribution and vertically in a uniform manner (Arndt et al., 1997). Point sources and volcanoes (important in this area) are treated as elevated sources and area emissions as surface sources. Each emission puff is modelled for up to five days, or when they leave the model domain. An example of modelled S deposition is given in Fig. 6.5.

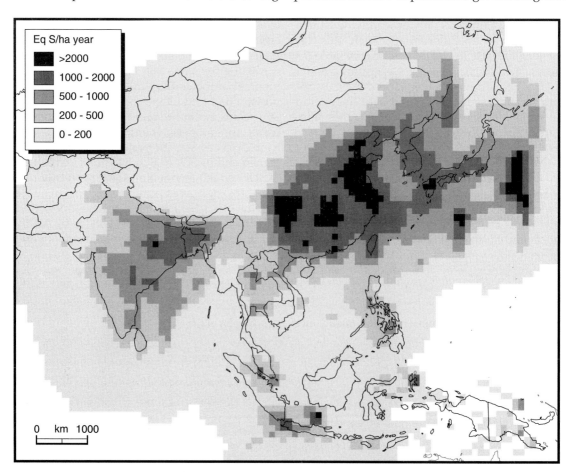

FIGURE 6.5 Sulphur deposition to Asia (in eq S/ha/yr) modelled using ATMOS model (Source: Arndt et al., 1997)

The application of ATMOS highlighted the importance of activities such as biomass burning and the role of volcanoes (contributing about 30 per cent of S deposition in Japan and 50 per cent in Indonesia), as well as the more predictable sources such as heavy industry and power generation. The authors point out, however, that a full evaluation of the model's performance was constrained by a lack of good measurement networks across the region (see Chapter 2). The development and extension of these networks has been the focus of recent activity.

6.3 Global air quality

6.3.1 Modelling the global atmosphere – linking composition to climate

6.3.1.1 Global climate models

As early as the mid-1950s, John von Neumann recognized that computers would be ideally suited to modelling the global climate, exploiting their ability to represent the physics of the atmosphere in mathematical terms (Kutzbach, 1985). By 1963, Smagorinsky had carried out the first comprehensive general circulation experiments. General circulation (dynamical) models, first of the atmosphere alone, then of coupled ocean and atmosphere and now moving towards ocean-atmosphere-biosphere models, have become the lynchpins of our efforts to understand how the climate system works. How human activities are perturbing this system and the changes this might bring in the future are key questions for us all. Climate models are conventionally tested by their abilities to reproduce the present day climate and patterns of climate change over the period of instrumental records (e.g. Mitchell *et al.*, 1995). It can also be helpful, however, to extend the timescale and compare modelled climates with climatic reconstructions based on palaeoenvironmental data for particular time periods in the geological past. Such an approach has benefits both for the modelling community and for those collecting and attempting to synthesize palaeoenvironmental records. Although palaeoclimate models have been developed for pre-Quaternary periods, it is the late Quaternary and Holocene (encompassing the last glacial maximum and the warming into the present interglacial, ca. the last 25,000 years), which has been the focus of most attention.

Climate models differ from weather prediction models because they have to run for decades or centuries rather than hours or days. In practical terms, therefore, climate models will have coarser levels of detail, although extra computing power has allowed the spatial resolution of climate models to improve. Early models had large grid sizes (often $8° \times 10°$) meaning that features such as surface topography and continental outlines were crudely represented. Output from these models suggests that North and South America have become disconnected and the Mediterranean landlocked! The finer resolution of recent models ($2° \times 3°$) corrects these geographical anomalies. There are, however, many key processes in the atmosphere (e.g. most cloud processes) that operate at the sub-grid scale and have to be parameterized in these models. General circulation models (GCMs) have become more and more complex (see above). Early models (AGCMs) dealt only with the atmosphere and its interaction with the land surface and cryosphere through albedo effects. The next step was to couple the atmosphere to a slab ocean, a shallow layer of water (usually about 50m deep) which allowed some, prescribed, ocean heat transport. The most complex physical models are coupled atmosphere-ocean general circulation models (AOGCMs). The HadCM2 and HadCM3 models from the UK Meteorological Office Hadley Centre are of this type (see their website http://www.met-office.gov.uk/ research/ hadleycentre/models/modeltypes. html).

Given the very different response times of the oceans and the atmosphere, bringing these two systems together is a complex process, but such models do allow better exploration of the important links between climate and the ocean circulation such as the North Atlantic Conveyor and the associated Gulf Stream.

Table 6.5 shows a selection of 16 global climate models that were used in the IPCC (2001) Third Assessment Report. The table presents their equilibrium climate sensitivities, defined as the temperature response of the global climate system to the radiative forcing (see Chapter 3) from a doubling in CO_2 concentrations, when steady state has been approached.

While relatively simple models can be used to assess the response of climate to factors such as changing orbital parameters (affecting the distribution of incoming solar radiation) and surface boundary conditions (e.g. more extensive ice sheets), models that look at changes in radiative forcing have to be able to represent other aspects of the composition of the atmosphere. To this end, aspects of global atmospheric chemistry models (see below) are increasingly being incorporated into climate models. All radiatively active gases with a positive forcing were initially represented by CO_2,

then it became apparent that aerosols, in particular sulphate aerosols, had an important part to play as a negative forcing (see Mitchell et al., 1995). The evolution of climate models to include more chemistry can be tracked through the various reports of the IPCC (1990, 1996, 2001). It has also become apparent that any attempt to look at the long-term future for atmospheric CO_2 levels has to consider the global carbon cycle. At the time of the IPCC Second Assessment (1995) this work was in its infancy, but had made considerable progress by the publication of the IPCC Third Assessment in 2001.

While it is hoped that increasing the complexity of models also increases their realism, it is not without its costs. Decisions have to be made about horizontal scale, numbers of vertical layers (ocean and atmosphere), which processes to represent explicitly and those that will be parameterized. Complex models take

TABLE 6.5 A selection of global climate models, taken from the Intergovernmental Panel on Climate Change (2001) Third Assessment Report, and their equilibrium climate sensitivities

Model	Equilibrium climate sensitivity, K	Reference
BMRC	2.2	Colman and McAvaney (1995)
CCSR/NIES	3.6	Emori et al. (1999)
CCSR/NIES2	5.1	Nozawa et al. (2001)
CGCM1	3.5	Boer et al. (1992)
CSIRO	4.3	Watterson et al. (1998)
CSM	2.1	Meehl et al. (2000)
ECHAM3	3.2	Cubasch et al. (1992)
GFDL_R15_a	3.7	Manabe et al. (1991)
GFDL_R30_c	3.4	Knutson et al. (1999)
GISS	3.1	Yao and Del Genio (1999)
HadCM2	4.1	Senior and Mitchell (2000)
HadCM3	3.3	Williams et al. (2001)
IPSL_CM2	3.6	Ramstein et al. (1998)
MRI1	4.8	Noda et al. (1999)
MRI2	2.0	Yukimoto et al. (2001)
DOE PCM	2.1	Washington et al. (2000)

more time (and data) to build and more time to run. Both the inputs and, perhaps more importantly, the outputs from complex models are more difficult to interpret. The sheer computational costs of the most complex models (e.g. AOGCMs) will limit the number of times they can be run and how long they can be run for. The relative merits of simple and complex climate models have been discussed by Houghton *et al.* (1997). There are considerable advantages to having access to both simple and complex models, as the outputs can be compared and key processes identified. If a simpler model can capture the essential behaviour of the system, then this may be a better tool for many purposes. This issue is explored below in relation to tropospheric chemistry models.

6.3.1.2 Palaeoclimate models

Running an AGCM for a past climate requires boundary conditions to be set. These include: incoming insolation, atmospheric composition (aerosols and greenhouse gas concentrations), land surface albedo and roughness, extent and height of ice cover over land and oceans and prescription of sea surface temperatures (SSTs). Many of these values will be set based on information such as the gas and dust content of ice cores (atmospheric composition), the isotopic composition of marine organisms (SSTs) and geomorphological evidence for ice limits (extent of ice cover). Model outputs can then be compared with climate reconstructions based on a range of proxy data (e.g. temperature data from pollen evidence, moisture availability from lake levels). The basic framework for this form of data–model comparison was set out by the COHMAP (Cooperative Holocene Mapping Project) group. The COHMAP group used only one climate model, the NCAR Community Climate Model (CCM). The version used for COHMAP had prescribed CO_2 and aerosol loading and for the last glacial maximum took its sea surface temperatures, sea and land ice extents from CLIMAP (CLIMAP, 1981). The results of the COHMAP experiments have been reported extensively (COHMAP, 1988; Wright *et al.*, 1993) and the maps comparing model climate and palaeo-data appear in

many text books. A more recent exercise, the Palaeoclimate Modelling Intercomparison Project (PMIP), involved 18 climate models, all forced by the same boundary conditions and tested against a variety of palaeoenvironmental data sets. One of the key aims of PMIP was to establish how much of modelled climate variability was actually model dependent.

The COHMAP experiments and many that have followed, were able to reproduce the broad patterns of climatic variability indicated by the data, but not the detail. Two particular details were noted: that the models could not generate cold enough temperatures at the last glacial maximum (18 k ^{14}C yr BP, 21 k BP) and that they were not able to mimic the strong increase in summer monsoon rainfall across Africa in the early Holocene (about 9 k ^{14}C yr BP) (Street-Perrott, 1991; Kohfeld and Harrison, 2000). It appeared that the use of the CLIMAP SSTs (widely recognized as underestimating the lowering of temperature in the tropical oceans at the LGM) or even a slab ocean (Manabe and Broccoli, 1985; Mitchell *et al.*, 1988) might be responsible. Vettoretti *et al.* (1998) report the use of an AGCM with and without coupling to mixed layer ocean and thermodynamic sea-ice modules to calculate SSTs and sea-ice cover for 6000 BP. Their results show that the coupled version of the model was better able to reproduce observed moisture balance changes over Africa than the version using fixed SSTs, but that the magnitude of the changes was still not as large as that indicated by the palaeo-data. It appears that successful modelling of the early Holocene changes in the African monsoon may require the inclusion of land-surface feedbacks (from lakes and wetlands, vegetation change) (Kohfeld and Harrison, 2000).

Until recently, the application of coupled ocean-atmosphere models to past climates has been very restricted as it has not been possible to run the long simulations required. Bush and Philander (1999) report the results of a decadal simulation of an AOGCM run for 21 k BP. There are many similarities between their simulation and those published previously for the last glacial maximum, such as anticyclonic flow around ice sheets, intensified Atlantic storm tracks and a cooling of the Northern Hemisphere continents of 10–15°C. Bush and Philander's model ocean can, however, respond to changes in the atmosphere. Over the North Atlantic, this lowers SSTs by 9°C and changes the orientation of the Gulf Stream to be more zonal (i.e. across the ocean, rather than penetrating as far northward as it does now). Significantly their model produced a cooling of 4–6°C in tropical SSTs, which together with a 20 per cent reduction in water vapour in the tropical atmosphere, resulted in a cooling in the tropics not captured in other models. The modelled tropical temperatures were much closer to those indicated by terrestrial tropical proxy data. The authors note that SST changes in the tropical Pacific had the largest climatic impact, further emphasizing the importance of this area to the global climate system.

Some of the latest generation of AOGCMs have been run for long integration periods. HadCM2, HadCM3 and the Geophysical Fluid Dynamics Laboratory (GFDL) model (Stouffer *et al.*, 1994) have all been allowed to run for a 1000 years or more. In these cases the models have not been 'forced' by changes in external factors, such as changes in greenhouse gas concentrations, but the focus has been on assessing whether the models can capture the internal, natural variability of the global climate system. Collins *et al.* (2001) report the results of a 1000 year integration of HadCM3. HadCM3 is seen as a significant improvement on its predecessors, as unlike many coupled ocean-atmosphere models, it does not require any flux adjustment (of heat or salinity) to prevent drift when modelling the present day climate. The ability to model internal climate dynamics is important if we are to be able to identify successfully anthropogenically induced climate change. If such models can also reproduce variability on annual to decadal timescales (such as the El Niño–Southern Oscillation (ENSO) and the North Atlantic Oscillation (NAO)), then we can have more confidence in using such models to forecast future events and their consequences.

The severe global impacts of major ENSO events are now well recognized. Broad agreement was found between modelled and measured surface temperatures for the period since 1851, although there were regional variations in model performance. Models were able to reproduce ENSO with an irregular three to four year cycle, but showed too much variability in the West Pacific Warm pool. Northern Hemisphere variability in the NAO was also simulated. Over longer time scales, HadCM3 was also able to simulate the Pacific Decadal Oscillation. In terms of trends through time, it appeared that the recent upward trend in the NAO index could not be explained by natural variability in the climate system, nor could the prolonged upward trend in global temperatures. In both cases, external forcing factors seem to have played a part.

6.3.2 Global air quality models

While Svante Arrhenius first quantified the effects of CO_2 and water vapour on the temperature of the earth's surface in the late nineteenth century (see Chapter 3) and there was anecdotal evidence of the impact of volcanic aerosols on climate from the Roman period (Charlson, 1997), an appreciation of the complex and close relationship between the chemistry of the atmosphere and the climate system has really only developed over the last 30 years (see Mitchell, 1970). This has come from an understanding that the radiative properties of the atmosphere are fundamentally affected by its chemistry and that human activities have changed (and are changing) atmospheric chemistry. Our appreciation of the extent to which the composition of the global atmosphere has changed since the early nineteenth century has come from the study of ice cores and limited direct measurements (see Chapters 2 and 3), but has also relied heavily on the development of global tropospheric chemistry transport models (CTMs). In common with climate models, these have developed through time from 1–, to 2- and now usually 3-D models.

Key challenges are that many of the radiatively active trace gases, as well as aerosols, have quite short atmospheric lifetimes and are actually regional in their distribution. As a result, the IPCC (1995) identified that only synoptic, 3-D CTMs would be able to operate on the spatial and temporal scales required for assessing the climatic effects of atmospheric chemistry. The most recent IPCC report (IPCC, 2001) has identified the following groups of radiatively important gases which depend on chemistry:

Direct GHG: ozone (O_3); methane (CH_4); nitrous oxide (N_2O); hydrofluorocarbons (HFCs) such as HFC-134a; perfluorocarbons (PFCs) such as carbon tetrafluoride (CF_4); sulphur hexafluoride (SF_6).

Indirect GHG: carbon monoxide (CO); volatile organic compounds (VOCs); oxides of nitrogen (NO_x).

The global, annual mean radiative forcings due to a number of agents for the period from pre-industrial times to the present day are shown in Fig. 3.11 on p. 51.

The radiative effects of the indirect GHG occur through their linkages to O_3, CH_4 and hydroxyl radical (OH) concentrations. Changes in OH are significant because of its role as an oxidant and the part it plays in the atmosphere's overall ability to clean itself. These gases contrast with the best known greenhouse gas, CO_2, whose lifetime in the lower atmosphere is not significantly affected by atmospheric chemistry. IPCC (2001) also highlights the positive radiative forcing of compounds which are probably better known for their role in the depletion of stratospheric ozone: CFCs, HCFCs, halons.

Since the papers of Fishman and colleagues (e.g. Fishman *et al.*, 1979), tropospheric ozone has been a major focus of attention given its importance both photochemically and climatically. Direct measurements and modelling work have shown the long-term upward trends in tropospheric ozone concentrations, particularly in the northern mid-latitudes (e.g. Hough and Derwent, 1990). Given the complexities of modelling the production and loss of tropospheric ozone, assessments of CTMs are often made in relation to this pollutant (e.g. IPCC, 1995). The IPCC 2001 report included a comparison of 14 models (see Table 6.6), capable of predicting O_3 and OH concentrations both for the present day and the future based in IPCC scenarios. These models, all of which included

3-D synoptic meteorology and representations of the chemistry of NOx, mon-methane VOCs and CO, were seen to be a major advance on earlier modelling efforts. The model projections through to 2100 all indicate major changes in tropospheric chemistry. The outputs from the models are summarized in Fig. 6.6.

All IPCC scenarios except one (B1) resulted in lower levels of OH, increasing the atmospheric lifetime of CH_4 from 8.4 to 10 years, and more persistent HFCs. As a result, both HFCs and CH_4 would have a longer period of positive forcing. All scenarios except B1 (global shift to service and information-based economies, global introduction of clean and efficient technologies) also showed an increase in tropospheric ozone concentrations. In two scenarios, ozone was modelled to increase by 60 per cent between 2000 and 2100, more than twice the change from the pre-industrial period to present. The report noted that non-CO_2 GHG are expected to increase substantially in the future and to contribute a sizable fraction to increased radiative forcing. Large increases in CO and O_3 concentrations downwind of major urban and industrial areas are expected to occur, with the impacts being intercontinental in scale. As described in previous chapters, many of these gases are implicated in other forms of pollution from urban to regional, with impacts on health and ecosystems.

Although the CTMs used in the IPCC assessment represent some of the latest generation, it was noted that there were important elements *not* included:

- the effects of the recovery of stratospheric ozone over the next century;
- no coupling of chemical change in the atmosphere to global biogeochemical cycles;
- no full coupling of chemical and climate models.

It is now recognized that, after carbon dioxide, methane and tropospheric ozone are the second and third most important greenhouse gases (IPCC, 2001). This recognition has been underpinned in large part by global air quality models.

Methane is a strongly radiatively-active trace gas emitted by both anthropogenic and natural

TABLE 6.6 Global chemistry-transport models contributing to the assessment of future methane and ozone burdens in the Intergovernmental Panel on Climate Change (2001) Third Assessment Report

Model	Type of model	References
GISS	Eulerian grid	Hansen et al. (1997)
HGEO	Eulerian grid	Bey et al. (2001)
HGIS	Eulerian grid	Mickley et al. (1999)
IASB	Eulerian grid	Muller and Brasseur (1995)
KNMI	Eulerian grid	Jeuken et al. (1999)
MOZ1	Eulerian grid	Brasseur et al. (1998a)
MOZ2	Eulerian grid	Brasseur et al. (1998b)
MPIC	Eulerian grid	Crutzen et al. (1999)
UCI	Eulerian grid	Hannegan et al. (1998)
UIO	Eulerian grid	Berntsen and Isaksen (1997)
UIO2	Eulerian grid	Sundet (1997)
UKMO	Lagrangian	Collins et al. (1997)
ULAQ	Eulerian grid	Pitari et al. (1997)
UCAM	Eulerian grid	Law et al. (1998)

processes (see Chapter 3). Methane has one major atmospheric removal process through its chemical reaction with the highly reactive hydroxyl radical (OH) which acts as the major cleansing reagent for the troposphere or lower atmosphere. OH oxidation gives methane an atmospheric lifetime of about nine years.

The tropospheric distribution of the hydroxyl radicals is maintained by the sunlight-driven photolysis of ozone which provides a continuous supply during daylight (see Chapter 4). Reaction with methane and carbon monoxide provide the main loss reaction for hydroxyl radicals. However, these reactions are only temporary sinks for OH because they generate hydroperoxy (HO_2) radicals which are immediately recycled back to OH through reactions with nitric oxide and ozone. There is therefore a continuous supply and recycling of OH which explains why it can act as such an efficient cleansing agent in the lower atmosphere. These fast chemical reactions also act as sources and sinks for ozone so that, along with the growth in global methane concentrations, there has been a steady growth in global ozone concentrations due to human activities.

Tropospheric ozone, like methane, is a strongly radiatively-active trace gas whose distribution in the lower atmosphere or troposphere, is controlled by both natural processes and by human activities (see Chapters 3 and 4). The main natural ozone source in the lower atmosphere is by transport downwards from the stratosphere and the main human-influenced source processes are the fast chemical reactions that control the distribution and abundance of OH as detailed in the paragraphs above. The main ozone sinks are destruction at the earth's surface and sunlight-driven photolysis, the source of the steady supply of OH radicals driving the fast chemistry of the lower atmosphere. Ozone lifetimes vary markedly throughout the lower atmosphere and overall are about 20 to 30 days.

The global scale build-up of methane and ozone which contribute to global warming is thus driven by increasing anthropogenic emissions of methane, carbon monoxide and NO_x, as indicated

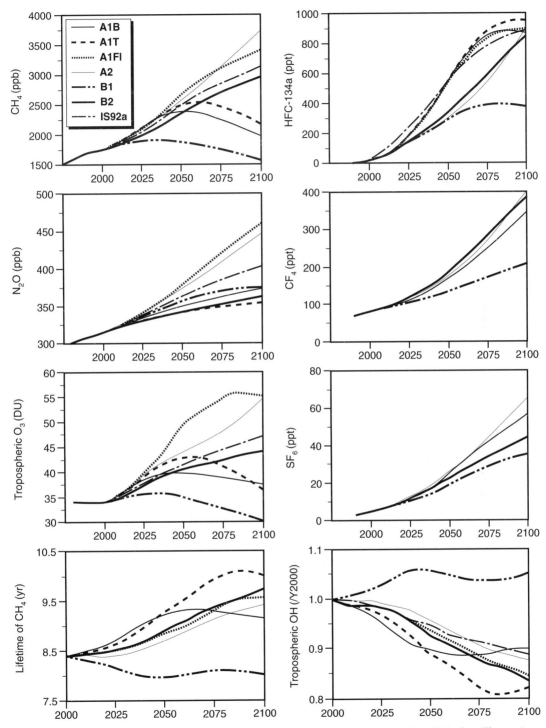

FIGURE 6.6 Atmospheric composition and properties predicted using the Six SRES Marker-Illustrative scenarios for anthropogenic emissions – abundances prior to 2000 are taken from observations (Source: IPCC, 2001)

by Fig. 6.1, where global warming impacts have been added in to the regional air quality impacts of acidification, eutrophication, ground level ozone and fine particle formation. This is the framework which underpins the application of global air quality models to the development of policies to control the global scale build-up of methane and tropospheric ozone.

In many ways, global air quality models closely resemble regional air quality models. The majority are Eulerian grid models and the majority employ similar chemistry, emission and deposition modules to those used in regional air quality models. One model, STOCHEM (Collins *et al.* 1997), adopts a Lagrangian air parcel approach in which 50,000 air parcels are moved with 3-D wind fields as an alternative to the Eulerian grid approach. A brief description of some of the global air quality models is given in Table 6.6. A comprehensive review of the tropospheric ozone budgets developed in these models is given by Collins *et al.* (2000). These models contributed to the assessment of the future methane and ozone burdens in the IPCC (2001) Third Assessment Report.

Global air quality models have added a new dimension to our understanding of global air pollution problems in their assessment of the impacts of emissions changes through to the year 2100. They have shown how the cumulative impact of all Northern Hemisphere emissions, not only those immediately upwind of each continent, may lead to ozone concentrations over much of the Northern Hemisphere mid-latitude region, which are double those currently observed. Since regional ozone episodes are superimposed on these background levels and build on them with local photochemical ozone formation, it may become increasingly difficult to achieve air quality guidelines for ozone in the future. This problem reaches across boundaries and couples emissions of NO_x on a regional and hemispheric scale. In the twenty-first century, a global perspective will be needed if regional air quality objectives are to be met. The impact of this potential degradation on regional air quality may change the balance of policy actions impacting on future emissions policies for greenhouse gases and regional pollution problems.

Pollution regulation

7.0 Introduction

Attempts to regulate emissions to air have seen a number of general trends since the earliest serious efforts to address this issue in the nineteenth century. These trends have included:

- a move from local action, to national and, most recently, international efforts to set emissions targets and goals for atmospheric composition;
- a change from permissive to mandatory legislation;
- the adoption of economic instruments;
- increasing public availability of environmental information (e.g. emissions to air, air quality data); and
- the setting of specified air quality standards with acceptable concentrations of particular pollutants.

The integration of regulatory frameworks across broader geographical areas (e.g. the EU, UNECE) has been matched by a change from addressing single issues (or pollutants) to integrated pollution control. The latter reflects the increasing understanding that the pollution 'climate' can be made up of a complex mix of pollutants whose combined effects may be more important than the impacts of any individual compound. As described in Chapters 3, 4 and 5, the same pollutants may play different roles at different scales and efforts to address issues at one scale must consider the effects in

other areas. This chapter will explore regulatory frameworks across three broad regions: the UK and western Europe, Eastern Europe and North America and address the role of international organizations, such as the United Nations, in setting the atmospheric pollution agenda. The final section will consider the increasing role of economic instruments, especially emissions trading.

Approaches to pollution control have been formalized under a variety of headings, integrating the expected level of technological intervention, cost considerations and overall approach. Some examples from Europe and North America are given in Table 7.1.

7.1 The United Kingdom and western Europe

In modern times, the widespread replacement of wood by coal as the major domestic fuel led to the first serious complaints about air quality and the first attempts at legislation. As early as 1273, a law was passed which attempted to prevent the burning of soft coal in English towns. It had no effect. Other, equally unsuccessful, attempts at local regulation continued to be introduced throughout the medieval and Tudor periods (Brimblecombe, 1987). The onset of the industrial revolution from the eighteenth century and the development of major industrial

TABLE 7.1 Approaches to air pollution control

Geographical area	Approach	Acronym
UK	Best Practicable Means	BPM
UK	Best Practicable Environmental Option	BPEO
UK/Europe	was Best Available Technology, now Best Available Techniques	BAT
UK	Best Available Techniques Not Entailing Excessive Cost	BATNEEC
Europe	Best Available Technology Not Entailing Excessive Cost	BATNEEC
USA	Best Available Control Technology	BACT
USA	Best Available Retrofit Technology	BART
USA	Maximum Available Control Technology	MACT

cities in the nineteenth century resulted in worsening air quality. Early legislation was driven by conditions in these urban areas which were both population centres and the focus of industrial activity (see Chapter 5). It was not until the nineteenth century that serious efforts at smoke abatement were introduced, starting with the Smoke Nuisance Abatement Act, 1853. Much of this legislation was in the context of laws relating to sanitary and public health conditions (Brimblecombe, 1987).

The first 'modern' piece of air quality legislation was the 1863 Alkali Act which required that 95 per cent of HCl emissions be stopped and the remainder diluted to the point where concentrations were no longer harmful. A second Alkali Act followed in 1874 and required that the approach of Best Practicable Means (BPM) be adopted (Ashby and Anderson, 1981). The Act specified particular noxious gases associated with so-called scheduled processes which were covered by the terms of the Act. A particularly important aspect of the legislation was the appointment of a national Alkali Inspectorate, the forerunners of the modern pollution inspectors working for the Environment Agency. Some of the major UK legislation since the 1863 Alkali Act is listed in Table 7.2.

In 1987, an Inspectorate of Pollution was created (HMIP) with the remit of dealing with pollution across all media (air, water, land) from scheduled processes based on the Best

Practicable Environmental Option (BPEO). The 1990 Environmental Protection Act implemented a two-pronged air pollution control system: Integrated Pollution Control (IPC) for large sources and Local Air Pollution Control (LAPC) which is the responsibility of local authorities. The Act gave HMIP responsibility for Part A processes (basically large industrial plant) in England and Wales under IPC, while local authorities had to deal with Part B processes under LAPC. In Scotland the Inspectorate dealt with Part A and Part B processes (IPC and LAPC). The BPM approach to pollution control, widely used previously, was replaced by the UK's version of BATNEEC. In 1995, the Environment Act set up integrated Environment Agencies for England + Wales and Scotland. These came into being in 1996 and kept their previous roles with respect to Part A and Part B processes. IPC and LAPC were introduced into Northern Ireland in 1998. The 1999 Pollution Prevention and Control Act replaced IPC and LPAC with Integrated Pollution Prevention and Control for part A processes and Local Authority Pollution Prevention and Control for part B processes. This followed the pattern set out in 1996 by an EU directive (see below).

Details of UK air quality legislation and consultation documents associated with policy development are available through the DEFRA website (http://www.defra.gov.uk/environment/airquality). Responsibility for

TABLE 7.2 Major UK legislation related to air pollution, since 1863

Year	Legislation
1863	Alkali Act
1874	2nd Alkali Act – national Alkali Inspectorate
1956	Clean Air Act
1968	Clean Air Act
1974	Control of Pollution Act
1987	Her Majesty's Inspectorate of Pollution created
1990	Environmental Protection Act, including Local Air Pollution Control (LAPC) and IPC (see below)
1991	Integrated Pollution Control (IPC) Setting up of Expert Panel on Air Quality Standards (EPAQS)
1993	Clean Air Act
1995	Environment Act – established new Environment Agencies and the requirement for local authorities to undertake Local Air Quality Management and declare Air Quality Management Areas where necessary.
1997	National Air Quality Strategy
1999	Pollution Prevention and Control Act. (Integrated Pollution Prevention and Control (IPPC) – first stage of implementation.)
2000	Air Quality Strategy revised
2003	Air Quality Strategy addendum

environment is devolved across the home nations of the UK through the Scottish Executive and the National Assembly for Wales and the Northern Ireland Assembly. Further information is available through the environment agencies: the Environment Agency for England and Wales (http://www.environment-agency.gov.uk); the Scottish Environment Protection Agency (http://www.sepa.org.uk) and the Northern Ireland Environment and Heritage Service (http://www.ehsni.gov.uk).

7.1.1 The European Union

When the Treaty of Rome was signed in 1957 setting up the European Economic Community (EEC), there was no direct reference to the environment as a community issue. The name was later changed to European Community (EC). The Community's early Environment Action programmes (starting in 1973) were couched in terms of reducing competition and encouraging fair trade. The environment, however, became an increasingly important element in European legislation. The Single European Act (1986) provided the legal basis for the EC to adopt environmental legislation (Gillies, 1999), setting out the essential principles for addressing pollution issues within the Community (e.g. the polluter pays). The Treaty on European Union of 1992 (the Maastricht Treaty) explicitly required community policies to consider environmental protection. The Maastricht Treaty also resulted in the adoption of the name European Union (EU). Within the EU, the European Commission is responsible for drawing up the Environment Action Programme and drafting environmental legislation (see below). In 1990, a European Environment Agency was established; it is based in Copenhagen and started work in 1994. Since

1997, the EU has had an explicit goal of integrating environmental concerns into areas of policy (e.g. transport, agriculture, energy) (see http://www.europa.eu.int/scadplus/leg/en/l vb/ l28066.htm). This is regarded as being particularly important for dealing with issues such as climate change.

The European Commission develops directives which are then incorporated into, and achieved through, national legislation by the member states. Some of these directives are issued in the form of framework directives. In some cases the European Commission signs international protocols (such as the 1990 revision of the Montreal Protocol and the 1992 Climate Change Convention – see Section 7.4.3) on behalf of the member states. It is the Commission's responsibility to ensure that the member states comply with the EC's environ-

mental laws and, if necessary, it can apply to take a member country before the European Court of Justice.

The EC approach to air quality was always to set specific acceptable standards and was different from the BPM approach then used in the UK. The first air quality directive related to SO_2 and smoke concentrations in urban areas (80/779/EEC). Through the 1980s, the EC was often seen as setting the pace in terms of environmental legislation, particularly with regard to air pollution, with the UK following reluctantly behind. There has now been a convergence of approaches. Some major EU directives relating to air quality are listed in Table 7.3. Further details on the EU's policy relating to air pollution issues are available at http://www.europa.eu.int/comm/environment/air/facts_en.htm.

TABLE 7.3 Major EU directives relating to air pollution

Directive	Topic
80/779/EEC	Standards for SO_2 and suspended particulates in air
81/462/EEC	Decision on the Geneva Convention on Long Range Transboundary Air Pollution
82/884/EEC	Set limit value of 2 $\mu g\ m^{-3}$ lead as annual mean concentration
84/360/EEC	Framework directive to prevent and reduce air pollution from industrial plants
85/203/EEC	Directive establishing limit value for NO_2 and lower guide values for special protection zones
87/217/EEC	Directive controlling asbestos in air, water and land from point sources
88/609/EEC	Large Combustion Plant Directive – emissions reduction targets for SO_2, NO_2 and dust
89/369/EEC	Directive setting limit values and operating requirements for new waste incinerators
89/429/EEC	Directive setting limits for existing incinerators
92/72/EEC	Directive on Ozone – monitoring, information and public alerts
94/67/EC	Directive on hazardous waste incineration
94/63/EC	Directive to reduce VOC emissions from evaporative fuel losses
96/62/EC	Framework Directive on Ambient Air Quality and Management
96/61/EC	Integrated Pollution Prevention and Control Directive
99/13/EC	VOC Solvents Directive
99/32/EC	Sulphur content of liquid fuels
2001/80/EC	Large Combustion Plant Directive (revised)
2001/81/EC	National Emissions Ceiling Directive

The 1996 Framework Directive on Ambient Air Quality and Management is designed to provide a comprehensive strategy for air quality management in the member states. It has four strands within it relating to: emissions from stationary sources, mobile sources, Air Quality Strategy and product controls (including ozone depleting substances). Although goals and criteria are set out by the EU, it is the individual member countries which have the responsibility to measure and assess air quality and to implement the directive. So called 'Daughter' directives will replace existing directives relating to air quality standards: 80/779/EEC, 82/884/EEC and 85/203/EEC (see Table 7.3). There is also a whole suite of EC directives relating to emissions from mobile sources (e.g. particulates from diesels, CO, hydrocarbons and NO_x) and the composition of liquid fuels, including petrol and diesel (e.g. lead content, sulphur content). All of these are subsumed within the scope of the framework directive. For details see http://www.europa.eu.int/comm/environment/air/ambient.htm.

In 1999, the Council adopted specific limit values for SO_2, NO_x, particulates and Pb. Further daughter directives in 2000 and 2002 addressed benzene, CO and ozone respectively. A 4th daughter directive relating to heavy metals and PAHs is in preparation. The revision of the Large Combustion Plant Directive (LCPD) and the new National Emissions Ceiling Directive (NECD) in 2001 saw significant new emissions targets and further controls on emissions from point sources.

The EU has played an active role in issues relating to emissions of greenhouse gases (associated with climate change) and of ozone depleting substances (see Chapter 3). The EC introduced monitoring of greenhouse gas emissions in 1993 (93/389/EEC). In 1999, monitoring was extended to greenhouse gases not covered by the Montreal Protocol (see Section 7.4.2) and member states were obliged to draw up plans to stabilize CO_2 emissions at 1990 levels by 2000. The EU ratified the UN Framework Convention on Climate Change (UNFCCC) (see Section 7.4.3) in 1994 and in 1998, the EC signed the Kyoto Protocol to the FCCC committing the member countries to reducing greenhouse gas emissions to 8 per cent below 1990 levels by 2008–12. In spite of these commitments, emissions of CO_2 have been increasing since 1994 in many member states. A strategy was clearly needed to enable the community to meet its commitments under the Kyoto Protocol and Buenos Aires action plan (see Section 7.4.3). In 2003, the EU Emissions Trading Directive was agreed (2003/87/EC) which will require installations covered by the directive to have a greenhouse gas emissions trading permit. This then allows them to trade any surplus allowance. The directive will come into force in 2005.

A Council Regulation on Substances that deplete the ozone layer (EC No. 3093/94) was introduced in 1994. The Commission is currently seeking approval to accept amendments to the Montreal Protocol agreed in 1997 (see Section 7.4.2).

Details of the EU's environmental policies can be obtained through http://www.europa.eu.int. scadplus/leg/en/s15000.htm.

7.1.2 Current air quality standards in the UK

In 1997, the UK government adopted a National Air Quality Strategy which was required under the terms of the 1995 Environment Act. The strategy set specific standards for acceptable concentrations of a range of air pollutants and target dates for meeting these standards. The strategy takes into account both health and more general environmental effects, as well as considering the costs and benefits of the abatement measures which would be required, together with the broader socio-economic issues. It was planned that the strategy should be subject to periodic review. The results of the first review were published in 1999 (DETR, 1999) and a revised strategy published in 2000 (DETR, 2000). An addendum to the strategy was introduced in 2003 (DEFRA, 2003) which set tighter targets for benzene, particles and CO and set a new objective for PAHs (polycyclic aromatic hydrocarbons). There has been an increasing tendency to set different targets for different parts of the UK (through the devolved administrations) reflecting different local conditions. The revised UK air quality standards are set out in Table 7.4.

Objectives were also set for the protection of vegetation and ecosystems with respect to NO_2 and SO_2. For NO_2 the objective was an annual mean of 15.7 ppb (30 μg m^{-3}) by 31.12.2000. For SO_2 the objective (annual and winter) was 7 ppb (20 μg m^{-3}) by 31.12.2000.

Air quality data are also reported in terms of bands: low, moderate, high and very high, based on the known health effects of each pollutant. The bands are related to thresholds: 1.) the standard threshold as set out in the UK's Air Quality Standard; 2.) EC Information threshold and 3.) the EC Alert level. These are summarized in Table 7.5.

TABLE 7.4 UK air quality standards under the revised Air Quality Strategy 2000 and its amendments 2003

Pollutant	Standard concentration	measured as	Year to be achieved (all 31 December)
Benzene	1.54 ppb (England & Wales) 1 ppb (Scotland & NI)	running annual mean	2010
1,3-butadiene	1 ppb	running annual mean	2003
Carbon monoxide	8.6 ppm	running 8-hour mean	2003
Lead	0.5 μg m^{-3} based on WHO guidelines	annual mean	2004
	0.25 μg m^{-3}	annual mean	2008
Nitrogen dioxide	150 ppb – not to be exceeded more than 18 times a year	1-hour mean	2005
	21 ppb	annual mean	2005
Ozone	50 ppb – not to be exceeded more than 10 times a year	running 8-hour mean	2005
Fine particles (PM$_{10}$)	50 μg m^{-3} – not to be exceeded more than 7 times a year (UK except London)	running 24-hour mean	2010
	20 μg m^{-3} (England*, Wales, NI)	annual mean	2010
	18 μg m^{-3} (Scotland)	annual mean	2010
Sulphur dioxide	100 ppb – not to be exceeded more than 35 times a year	15-minute mean	2005
	47 ppb	24-hour mean	2004
Polycyclic aromatic hydrocarbons	0.25 ng m^{-3} B[a]P (GB)	annual mean	2010

* except London (where the annual mean limit is 23 μg m^{-3})

TABLE 7.5 Air quality bands used in the UK

Pollutant	Low	Moderate	High	Very high
SO_2 (ppb, 15-min av.)	< 100	100–199	200–399	400+
PM_{10} ($\mu g\ m^{-3}$, 24-hr av.)	< 50	50–74	75–99	100+
NO_2 (ppb, hourly av.)	< 150	150–299	300–399	400+
O_3 (ppb, running mean or hourly mean)	< 50	50–89	80–179	180+
CO (ppm, 8-hr av.)	< 10	10–14.9	15–19.9	20+

In 1991, the Expert Panel on Air Quality Standards (EPAQS) was set up to provide the UK government with independent advice on identifying levels of pollution at which minimal or no health effects would occur. So far, EPAQS have set standards for nine pollutants: benzene, ozone, 1,3-butadiene, carbon monoxide, sulphur dioxide, particles, nitrogen dioxide, lead and polycyclic aromatic hydrocarbons (PAHs). All EPAQS reports are available from the DEFRA website (http://www.defra.environment.gov.uk/environment/airquality/aqs/index.htm). The standards recommended by EPAQS (Table 7.6) generally formed the basis for the National Air Quality Strategy, but in some cases these are derived from EU limit values (themselves based on WHO standards).

In 2002, a new group was established: the Air Quality Expert Group (AQEG). Their role is to provide scientific advice to the UK government on all aspects of air quality, but particularly in relation to those pollutants covered by the Air Quality Strategy.

7.2 Eastern Europe

7.2.1 The Former Soviet Union

The Socialist approach of collective ownership and central planning could have provided an effective approach to environmental protection and pollution control. In theory, the former USSR had some of the strictest environmental regulations in the world, but these were never implemented for reasons ranging from a lack of

TABLE 7.6 Air quality standards recommended by EPAQS

Pollutant	Year set	EPAQS standard
PAHs (based on benzo[a]pyrene)	1999	0.25 ng m^{-3} B[a]P annual average
Lead	1998	0.25 μg m^{-3} annual average
Nitrogen dioxide	1996	150 ppb hourly average
Particles (PM_{10})	1995	50 μg m^{-3} 24-hour running average
Sulphur dioxide	1995	100 ppb 15-minute average
Carbon monoxide	1994	10 ppm 8-hour running average
1,3-butadiene	1994 and 2002	1 ppb annual running average
Ozone	1994	50 ppb 8-hour running average
Benzene	1994	5 ppb annual running average, target 1 ppb annual average

capital to introduce the necessary abatement measures, to a lack of political will. The traditional goal was to achieve economic growth through industrial development which resulted in major concentrations of heavy industries in areas such as Siberia, Ukraine and south of the Ural Mountains. Smelting of high S copper-nickel ores (up to 25 per cent S) around Nickel (Kola Peninsula) and Noril'sk-Talnakh (Taymyr Peninsula) resulted in massive emissions of SO_2 – 2.3 million tonnes in 1995. The equipment used in these industries was often old and poorly maintained. Heavy industry (especially the ferrous and non-ferrous metal industries) and power generation (usually burning coal or heavy fuel oil) were the major sources of pollution, with SO_2 and particulates the main pollutants. Locally high concentrations of heavy metals and VOCs were also major problems (Shahgedanova and Burt, 1994). Relatively low levels of vehicle ownership, however, meant that emissions of NO_x and CO from this source were relatively low. Patterns of development meant that air quality was often poorer in small to medium size towns than in the largest cities, although this is now changing. The need for environmental regulation was not recognized centrally until the 1980s and there was very little environmental data available. Even in the early 1990s it was estimated that about one third of the industrial enterprises in Russia had no documentation on their emissions (Peterson, 1993).

The USSR made a first effort at controlling air pollution with a resolution passed in 1949 and the adoption of an Air Quality Management Strategy in 1951. Stringent air quality standards were set (covering more than 200 pollutants by the early 1980s) and monitoring systems were put in place. Some 24-hour air quality standards are listed in Table 7.7.

The locations of monitoring sites were determined by population (all cities with a population >100,000 had to have at least one monitoring site) (Peterson, 1993). Monitoring, however, was intermittent (i.e. for 20 to 30 minutes, two to four times per day) rather than continuous (Shahgedanova *et al.*, 1999) which has made comparisons with data from other west-

ern European or North American networks problematic.

The first national law on air pollution, the Air Quality Law, was passed in 1980. This aimed to set emission standards, while looking at environmental quality in the context of economic and social planning, but proved difficult to enforce. One of the problems was the very complex regulatory framework, with 15 agencies involved in monitoring and controlling air quality prior to 1988. The efficiency of the legislation was further undermined by the involvement of a number of powerful ministries (e.g. Ferrous and Non-ferrous metallurgy) which were responsible for both major sources of pollution and their regulation (Shahgedanova and Burt, 1994). Also in 1980, the Soviet Meteorological Service (Goskomgidromet) began collecting emissions data for the two broad categories of stationary sources and transport. Only a limited range of sources were actually included in these inventories and it appears that emissions across the full range of pollutants (e.g. SO_2, NO_x, CO_2) are likely to be underestimated both for the USSR and now for the Russian Federation (Peterson, 1993; Hill, 1999).

A major step forward came in 1988 when The USSR State Commission for the Protection of Nature (Goskompriroda) was set up to replace the previous chaotic administrative structure of environmental protection (Peterson, 1993). In 1989 Goskompriroda were responsible for producing the first comprehensive review of environmental conditions in the USSR (The State of the Environment in the USSR) which brought

TABLE 7.7 24-hour air quality standards for the USSR (Source: Shahgedanova and Burt, 1994)

Pollutant	24-hr standard
SO_2	50 µg m^{-3} (19 ppb)
Particulates	150 µg m^{-3}
CO	3000 µg m^{-3} (2610 ppb)
O_3	30 µg m^{-3} (15 ppb)
NO_2	40 µg m^{-3} (21 ppb)

the extent of the problems out into the open for the first time. A comprehensive programme for environmental amelioration was set out in the thirteenth five-year plan (1991–95), but the Soviet Union collapsed before it could be implemented. The former republics of the USSR now have their own legislative programmes to deal with environmental issues. A review of environmental conditions across the 12 nations of the Commonwealth of Independent States was published in 1996, but it has been noted that this report was less detailed than its 1989 predecessor and that data quality was a continuing problem (Shaw and Oldfield, 1998).

The Russian Federation, comprising the federal government and 89 sub-national units/territories, is still the largest country in the world. A new umbrella law relating to the environment was passed in 1991 and specific legislation relating to air pollution has been introduced. In 1997, overall responsibility for environmental management, monitoring and enforcement passed to the State Committee for Environmental Protection (Goskomekologiya), which had regional and local branches. This has now been disbanded. The institutional framework within the Federation is, however, very complex and inefficient as responsibilities are scattered and sometimes overlap (OECD, 1996). A United State System of Environmental Monitoring has been proposed with better quality assurance systems and more analysis and interpretation of data. The need for more modelling work has also been identified to assist in policy formulation. Some work has been done on this through a joint Russian/US EPA programme (Russia Air Management Project), which began in 1992 (see http://www.epa.gov/air/oaqps/ramp/index.html).

Pollution from large industrial sources, coal fired power stations and urban air quality remain major issues. Norilsk Nickel (see above) is still the largest single source of air pollution in Russia. High levels of air pollution are associated with respiratory health problems, with metals and benzene giving rise to excess cancers (special issue of *Environmental Science and Pollution Research* 2003, vol. 10 (1)). In spite of industrial decline, urban air quality is poor, largely due to increasing numbers of motor vehicles and the continuing use of leaded petrol. Health standards are regularly exceeded. In 1995, the Environment Ministry estimated that 66 per cent of Russia's population was breathing severely polluted air. The situation in Moscow has led to factory closures, restrictions on emissions from local oil refineries, restrictions on lorries during certain times of day and efforts to ban leaded petrol and to insist that cars have catalytic convertors.

The USSR signed the UN Convention on Long Range Transboundary Air Pollution (see below) and its two earliest protocols. The Russian Federation later signed the Oslo Protocol, but has not ratified it. It has not signed the Aarhus (heavy metals and POPs) or Gothenburg Protocols (see Table 7.12 on p. 150). Emissions of SO_2 and NO_x have declined, but this is mainly due to the marked decline in industrial production since the break up of the former Soviet Union and a switch from coal and fuel oil to natural gas (Hill, 1997).

The Russian Federation ratified the UN Framework Convention on Climate Change (1992) and signed the Kyoto Protocol (1997). It did not ratify Kyoto and it has been suggested that this was due to its status as one of the world's largest oil exporters. Russia agreed only to stabilize its CO_2 emissions at 1990 levels, which is a very lenient target compared with EU countries and indeed other Eastern European countries such as the Czech Republic (see below). The Russian Federation set out its position in its Third National Communication (2002) and ratification appeared unlikely. There does, however, seem to have been a change of heart. In May 2004, President Putin announced that Russia would 'rapidly move towards ratification' following complex negotiations with the EU. In November 2004 he signed the Federal law to ratify the protocol. This move is seen as key for the Kyoto Protocol (see section 7.4.3).

Emissions of CO_2 from Russia are considerable and, given the likely underestimation in the published inventories, may actually be equal to those of the USA (Hill, 1999). It is thought to be the third largest emitter of CO_2 in the world, of which gas is the largest source

(Marland *et al.*, 2003). Emissions of CO_2 have largely followed GDP (gross domestic product). As this fell between 1990 and 1996, so did CO_2 emissions. Overall, CO_2 emissions fell 28 per cent between 1992 and 2000, but are still 392 M tonnes C. Overall, greenhouse gas emissions fell until 1999 and then started to increase again. CO_2, CH_4 and N_2O have shown an overall downward trend, while emissions of HFCs, PFCs and SF_6 have increased. Emissions of greenhouse gases per unit GDP are high and higher now than in 1990. The Russian Federation is the world's third largest consumer of CFCs after China and South Korea.

In relation to stratospheric ozone depletion, the USSR signed the Vienna Convention and the Montreal Protocol (see below), but no protocols or amendments have been signed since 1992.

Some information on the Russian Federation is available through http://www.grid.ecoinfo.ru/webint_eng/start.htm.

7.2.2 The Czech Republic

As Czechoslovakia, the Czech and Slovak Republics formed part of the Soviet Bloc; this came to an end in 1989. The burning of soft, brown coal (with an S content of 1.5–2.4 per cent) and intensive development of metallurgical and chemical industries resulted in severe air pollution problems which first became apparent after World War I and intensified through the 1960s and 1970s. The worst affected area was northern Bohemia, part of the infamous 'Black Triangle', and the major pollutants of concern were SO_2 and particulates. In 1990, the region of north Bohemia was responsible for 42.6 per cent of the Czech Republic's SO_2 emissions. The communist regime ensured low cost energy which resulted in high energy consumption in inefficient industrial plants.

The constitution of the Czechoslovak Republic (1960) incorporated the principle of care for the human environment, with the federal government taking responsibility for international cooperation and coordination, developing the underpinning science and initiating legislation. Environmental laws were actually the same in the Czech and Slovak areas. The first federal law relating to air pollution was introduced in 1967 setting out areas of responsibility, monitoring requirements and scales of (low) fines (Hrbacek *et al.*, 1989). Further legislation relating to air pollution from vehicles and the protection of forests followed in 1975 and 1977. In 1981, Czechoslovakia signed the UNECE Convention on Long Range Transboundary Air Pollution. As in other eastern European countries, environmental issues came increasingly to the fore through the 1980s, although existing legislation proved ineffectual.

In the early 1990s, two factors came together which resulted in dramatic changes in emissions from the Czech Republic (and improved air quality). These were the implementation of new legislation and a sharp decline in industrial activity as the economy began to restructure after the end of the communist era. A Clean Air Act was introduced in 1991 (amended 1992) to the Czechoslovak Federal Republic, which was then adopted by the Czech Republic (Murley, 1995). The Clean Air Act defined regulated pollutants and emission limits for both stationary and mobile sources. New emission limits meet EU standards (discussed further below). The 'polluter pays' principle has been adopted and higher fines and more stringent penalties have been introduced for breaching limits. The state authorities now have the power to enforce controls, or close the offending plant. The 1991 act has subsequently been amended to tighten controls on vehicle emissions (1994) and stationary sources (1997). Public access to air quality data has been introduced, and an environmental yearbook has been published since 1993. In 1992, a smog regulation system was set up, with special protection zones identified. A national Air Quality Information System (ISKO) was established in the same year (http://www.chmi.cz/uoco/isko/schisko/schiskoe.html).

Some of the air quality standards introduced in 1991 are given in Table 7.8 (Czech Environmental Institute, 1999). These were adopted by the Czech Republic when it was established in 1993.

TABLE 7.8 Selected air quality standards for the Czech Republic (from 1991)

Pollutant	Annual	Daily	8-hour
SO_2	60 µg m^{-3} (22.5 ppb)	150 µg m^{-3} (56.25 ppb)	N/A
Suspended particulate matter	60 µg m^{-3}	150 mg m^{-3}	N/A
NO_2	80 µg m^{-3} (41.6 ppb)	100 µg m^{-3} (52 ppb)	N/A
CO	N/A	5000 µg m^{-3} (4350 ppb)	N/A
O_3			160 µg m^{-3} (80 ppb)

These standards are similar to those in operation elsewhere (see Tables 7.4, 7.10, 7.11).

There are three tiers of administration: the Ministry of the Environment (which directs the Czech Environmental Inspectorate), district offices and municipal authorities. The Ministry of the Environment deals with air quality monitoring and the provision of information (through the Czech Hydrometeorological Institute). The Environmental Inspectorate deals with emissions and compliance by large and medium sources, monitors compliance with smog control measures and monitors legal actions against polluters. In 1995, the Czech government approved the State Environmental Policy of the Czech Republic, which sets out their long-term priorities and principles of environmental protection for a wide range of issues including climate change. In 2001, it adopted a State Environmental Policy identifying high priority issues such as the country's high emissions of greenhouse gases.

Air quality data are available from the Czech Hydrometeorological Institute as annual or daily records (http://www.chmi.cz/uoco/oco_maine.html). Summaries of air pollution in the Czech Republic are also available through the Czech environmental yearbooks. These are available as hard copy or through the Internet at the Hydrometeorological Institute site or through the Phare website (see below).

The Czech Republic is one of a number of eastern and central European countries who joined the EU in 2004. The need for convergence between the environmental legislation in these countries and that in the EU was identified as a serious issue (Nath, 1997). As part of the process of European integration, the Czech Republic (and other EU accession countries) are members of the EU Phare Programme (http://www.europa.eu.int/comm/enlargement/pas/phare/intro.htm), which addresses environmental, health, education and other issues relevant to economic and political restructuring.

The Czech Republic hosts the website for the Phare Topic link on Air Quality (http://www.chmi.cz/uoco/isko/ptl). EU directives formed the basis for environmental legislation in the Czech Republic prior to accession. In 2002, a New Clean Air Act was introduced for health, vegetation and ecosystems using the EU limit values. Some of the standards set in 1991 (see Table 7.8) are still used, however, particularly for measures not covered by the EU (e.g. SPM).

The Czech Republic acceded to the UNFCCC in 1993 (see Section 7.4.3). Emissions of greenhouse gases fell sharply between 1990 and 1995 (by about 20 per cent), but this was associated with a period of economic decline and major restructuring. Growth in GDP resumed in 1994 and is now higher than at any time in the 1990s. Legislation relating to climate change is being harmonized with that of the EU. Emissions of greenhouse gases fell quite steeply through the early 1990s, increased in 1996/97 and have since fallen again. Only CH_4 has shown a consistent decline, with emissions of HFCs, PFC and SF_6 increasing since they were first recorded in 1995 (Ministry of the Environment, 2001). The burning of solid fuel (coal/coke) still contributes about 50 per cent of the country's energy supply, although consumption of brown coal fell by 49 per cent between 1990 and 1999. Solid fuel use is declining as gas is increasing. The Czech Republic signed the Kyoto Protocol in 1998, but projections of greenhouse gas emissions indicate that these may be higher (a high emission scenario) in 2020 than in 1999 even with measures to improve energy efficiency and the use of renewables. The Czech Hydrometeorological Institute hosts the National Centre for the UN Framework Convention on Climate Change and reports and data are available at http://www.chmi. cz/cc/acc/aindex.html.

The first controls on ozone depleting substances (as defined in the Montreal Protocol) were introduced in 1993 and extended in 1995 and 1996. The Czech Republic has ratified all the amendments to the protocol.

Although the Czech Republic lay within the Soviet Bloc until 1989 it has adopted a very different attitude towards air pollution legislation than the Russian Federation and most other parts of the former USSR. In part this reflects the different economic conditions in these 'economies in transition' (Missfeldt and Villavicenco, 2000). Industrial decline in the immediate post-communist period led to reductions in the emissions of most pollutants across eastern Europe. This decline in emissions was, however, often less than the decline in industrial production, reflecting the highly polluting nature of many industries. The failure of the economies of the former Soviet Union to resume growth has meant that environmental issues have become a low priority; there is little political will to address these issues or money to implement modern, clean technologies. In the Russian Federation, much of the progress of the last years of the communist era seems to have been lost. Elsewhere in central and eastern Europe new environmental policies have been developed and improved environmental standards enforced. This trend has been particularly pronounced in accession countries to the European Union.

7.3 North America

7.3.1 The United States of America

As in Europe, air pollution in the USA was initially viewed as a local problem. Individual cities passed legislation trying to control smoke in the late nineteenth century; state laws dealt with nothing other than smoke until 1947. The 1940s saw the emergence of smog as an air pollution issue in the Los Angeles Basin (see Chapter 5) and although its cause was not known, the California Air Pollution Control Act was passed in 1947. The relationship between motor vehicle emissions (NO_x and VOCs) and smog formation was identified in the early 1950s and in 1960 the California Motor Vehicle Pollution Control Act was passed which required new vehicles to have pollution control technology. Legislation in California led the way for subsequent action at the federal level.

Federal legislation on air pollution issues was stimulated by the spread of the smog problem and deaths in New York and Pennsylvania (among others). The first major piece of legislation was the 1955 Air Pollution Act, although this left regulation to state and local government. It was the Clean Air Act of 1963 that introduced the first, limited, federal powers to abate emissions seen as a health hazard. Some of the major legislation in the USA is listed in Table 7.9.

The 1970 Clean Air Act represented a marked major expansion in federal involvement in air quality management. It was a comprehensive federal law covering emissions from area, stationary and mobile sources and authorized the EPA (see below) to establish National Ambient Air Quality Standards (NAAQS). The Republican administration of Ronald Reagan did not reauthorize the Clean Air Act in 1981, so the 1970 act stayed in force. Urban air pollution continued to get worse and there was very severe photochemical pollution in the summer of 1988. The pressure for further action which this caused led to the Bush (Snr) administration introducing the Clean Air Act Amendments in 1990. These broadened the scope of federal

TABLE 7.9 Major federal air pollution legislation in the USA

Year	Legislation
1955	Air Pollution Act
1963	Clean Air Act
1965	Motor Vehicle Pollution Control Act
1967	Air Quality Act
1970	Clean Air Act (establishing National Ambient Air Quality Standards, NAAQS)
1977	Further Clean Air Amendments
1978	Ban on fluorocarbon gases in aerosol products
1982	Emissions Trading policy
1987	Ratified Montreal Protocol
1990	Clean Air Act Amendments
1991	Air Quality Agreement (USA and Canada) Rule giving right to buy and sell SO_2 emissions
1997	Clean Air Act revisions

regulations and attempted to use market forces and incentives to improve air quality; the system of emissions trading and banking, for example, was extended.

The US Environmental Protection Agency (EPA) was set up in 1970, the same year as the Clean Air Act, with a remit to set new standards and to intervene at the state level where necessary. The EPA took over responsibility – to 'protect human health and to safeguard the natural environment' – from a range of government departments and councils. Previously, the Department of Health and Welfare had been responsible for the National Air Pollution Control Administration. The EPA set up an Air Pollution Control office and now deals with atmospheric pollution issues through its Air and Radiation programme. It should be noted that the EPA continues to deal with releases/discharges to the different media (e.g. air, water) separately and does not apply the integrated approach to pollution control which has been adopted in Europe.

The setting of air quality standards has been a key aspect of air pollution policy in the USA and has more recently been adopted within the EU.

The 1963 Clean Air Act set up a programme to establish criteria for air pollution effects and the 1967 Air Quality Act required states to set air quality standards consistent with the federal criteria. The establishment of National Ambient Air Quality Standards (NAAQS) through the 1970 Clean Air Amendments marked the shift of responsibility for setting standards to the federal level, through the EPA. The EPA recognized a series of criteria pollutants (NO_2, SO_2, CO, ozone, PM_{10} and lead) for which they set primary standards relating to health, but also secondary standards relating to the environment (visibility, materials etc). Non-criteria pollutants fell into two categories – hazardous or toxic – with standards based on risk or occupational exposure levels. The NAAQS proved difficult to implement, so Amendments were introduced in 1977. These extended the deadline for compliance with the standards and introduced the idea of Prevention of Significant Deterioration (PSD). PSD allowed no deterioration in air quality in regions designated as Class I, identified as pristine (wilderness, National Parks) and then outlined a scale of allowable change through to Class III areas of industrial growth. The PSD programme required new sources to have BACT (see Table 7.1) and in some cases the application of BART.

The 1990 Clean Air Act Amendments marked a significant revision of previous legislation. The Act had 11 major provisions (or titles) addressing issues such as NAAQS, hazardous air pollutants, acid deposition, stratospheric ozone depletion, research and employment. It allowed for the establishment of interstate commissions on air pollution control to address regional issues and looked at the long-range transport of pollutants between the USA, Canada and Mexico. Important aspects of the 1990 Act were: the introduction of permits for emissions to air at the national level (previously administered at the state level), new powers of enforcement and setting a clear timetable for action through to 2010. Details of the Act are available from the Office of Air and Radiation (http://www.epa.gov/oar/oaq_caa.html). The NAAQS for criteria pollutants set out in the 1990 Act and in the 1997 revisions (for ozone and particulate matter) are given in Table 7.10.

Up to date information on standards is available through the Office of Air Quality Planning and Standards (http://www.epa.gov/airs/criteria. html).

The new (1997) standards for ozone and particulates were successfully challenged in the US Court of Appeal in 1999, in spite of support from the Clinton White House. This decision is itself the subject of an appeal. Also in 1997, the EPA proposed regional haze regulations to improve air quality and visual air quality in 150 natural areas (including many National Parks) across the USA. In 2000, the EPA introduced more stringent standards for vehicle emissions and lowered the permitted S content of fuels for cars and light goods vehicles.

In 2003, President George W. Bush put forward his 'Clear Skies' legislation which focuses on reducing emissions of SO_2, NO_x and Hg. The Act is currently under consideration.

The USA is a signatory to the Montreal Protocol on Substances that Deplete the Ozone Layer (see Section 7.4.2). To implement this

TABLE 7.10 USA NAAQS for criteria pollutants based on the 1990 and 1997 Clean Air Act and revisions

Pollutant	Standard type	Concentration	Measured as
CO	Primary	9 ppm (10 mg/m^{-3})	8-hour average
	Primary	3 ppm	1-hour average
NO_2	Primary & secondary	0.053 ppm (100 µg m^{-3})	Annual average
O_3	Primary & secondary	0.08 ppm (157 µg m^{-3})	8-hour average
Pb	Primary & secondary	1.5 µg m^{-3}	Quarterly average
PM_{10}	Primary & secondary	50 µg m^{-3}	Annual average
		150 µg m^{-3}	24-hour average based on 99th percentile of 24-hr measurements
$PM_{2.5}$	Primary & secondary	15 µg m^{-3}	Annual average
		65 µg m^{-3}	24-hour average based on 98th percentile of 24-hr measurements
SO_2	Primary	0.03 ppm (80 µg m^{-3}) 0.14 ppm (365 µg m^{-3}) 0.5 ppm (1300 µg m^{-3})	Annual average 24-hour average
	Secondary		3-hour average

protocol, the Clean Air Act Amendments (1990) were expanded to include monitoring of ozone depleting substances and a timetable for the phasing out of their production and consumption. Other regulations under the CAA cover issues such as identifying safe alternatives to known ozone depleting substances, issues relating to recycling and labelling. The most effective ozone depleting substances (e.g. CFCs, carbon tetrachloride and methyl chloroform) were all due to be phased out of production by 2002. Details of this timetable can be found on the EPA's website (http://www.epa.gov/ozone).

In contrast to its positive attitude towards protecting the stratospheric ozone layer, the USA has been slow to agree to reduce any emissions implicated in global climate change (see Chapter 3). The USA signed and ratified the UN Framework Convention on Climate Change (see Section 7.4.3), and has signed but not ratified the Kyoto Protocol to this convention. The Clinton administration introduced a Climate Change Action Plan in 1993, but the focus was on voluntary initiatives to reduce greenhouse gas emissions. There have also been efforts to encourage the expansion of renewable energy sources and to encourage energy efficiency. US emissions of greenhouse gases, however, continued to rise throughout the 1990s. By 2001, total greenhouse gas emissions were 13 per cent more than in 1990 (although they were slightly lower in 2001 than in 2000) (EPA, 2003) – the USA remains the world's largest emitter of greenhouse gases. The lack of participation in international efforts to reduce emissions has largely been due to a very strong lobbying campaign led by the Global Climate Coalition – a grouping of industrialists mainly associated with the oil industry, power generation and motor manufacturers. Attacks on the need for action have also been launched through the National Center for Policy Analysis. For a sceptical view of climate change see the NCPA website at http://www.eteam.ncpa.org/policy/Global_Warming/. There are signs that attitudes to global warming are changing in the USA, but these changes have yet to result in a new approach to climate change policy in the US Congress.

The change at the White House from the Clinton to Bush presidency resulted in a significant shift in policy towards the environment and the relationship between the EPA and the White House has been strained. This has led to some quite bitter public exchanges, particularly over climate change science. There is clear scientific support for human-induced climate change among the US science community (http://www.yosemite.epa.gov/oar/globalwarming.nsf/content/index.html), but this is not being carried through into policy at the federal level. It is notable, however, that individual states are taking action and agreements to reduce greenhouse gas emissions being made through the US EPA Climate Leaders initiative (http://www.epa.gov/climateleaders). The new Bush administration (2004) has promised to reduce emissions but still refuses to sign up to Kyoto.

7.3.2 Canada

In common with other federal systems of government, environmental legislation in Canada can be introduced at the federal or provincial (or indeed more local) level. Air pollution first became an issue in Canada in the 1920s when it became apparent that smelter operations in British Columbia were causing damage to vegetation in both Canada and the USA. In the 1940s, transboundary air pollution from the area around Detroit (USA) was recognized. The period after World War II saw increasing concerns about urban air quality. Acid deposition (see Chapter 4) is seen as a major environmental issue in eastern Canada, and the province of Ontario was the first area to implement air pollution legislation in 1958. Action at the federal level has occurred mainly since the 1970s. In 1970, a Federal Department of the Environment was set up (with equivalents at the provincial level); this is now known as Environment Canada. The first Federal Clean Air Act was passed in 1971 which gave the federal government the right to set emissions standards from stationary sources where human health or an international agreement were at risk (Barker and Barker, 1988). The act was amended in 1980. The Clean Air Act allowed for the setting of emissions standards from stationary and mobile sources, the setting of national ambient air quality standards and the regulation of the composition of fuels. These issues are now covered by the Canadian Environmental Protection Act, first passed in 1988 and revised in 1999.

National Ambient Air Quality Objectives (NAAQOs) are currently set for five pollutants (SO_2, NO_2, O_3, CO and total suspended particulates (TSP)), with objectives proposed for hydrogen sulphide and hydrogen fluoride. Objectives are set for different time periods and for three air quality ranges – desirable, acceptable and tolerable – with the limits set at the maximum of each range. The acceptable level is similar to the primary standard in the USA and is the standard adopted by most of the Canadian provinces. Objectives for the acceptable range are set out in Table 7.11. Full details of the NAAQOs can be obtained from http://www.hc-sc.gc.ca/hecs-sesc/air_quality/regulations.htm#3.

In 1998, a Canada-wide accord on environmental harmonization was introduced which aims to implement Canada Wide Standards (CWS). These harmonize environmental standards and bring in a common time frame for their implementation. CWS have been set for emissions of benzene, mercury, concentrations of very fine particles ($PM_{2.5}$) and ozone. The standard for O_3 is 65ppb as an 8-hour average (based on the fourth highest measurement over three years) by 2010. It is recognized that transboundary pollution from the USA may cause

problems in meeting this target to the extent that Ontario has just been set emissions reductions targets for NO_x and VOCs (as the main O_3 precursors). It will be assumed that any remaining exceedance of the standard can be attributed to the USA!

As described above, the transboundary nature of air pollution has long been recognized in Canada. The Eastern Canada Acid Rain Program, together with the US Acid Rain Program, formed the basis of the Canada–US Air Quality Agreement signed in 1991. This agreement has focused on setting targets for reductions in emissions of SO_2 and NO_x, but since 1997 there has been increasing emphasis on the need to control the precursors of ozone and particulate matter. Reports on the progress of the Air Quality Agreement are available from Environment Canada and through their website (http://www.ec.gc.ca/report_e.html). In 2000, an Ozone Annex to the Canada-US Air Quality Agreement was agreed to try to reduce transboundary low level ozone. In 2001, Canada produced an Interim Plan on particulates and ozone with more stringent emissions regulations for vehicles. Other strands have been to reduce VOCs from solvents, from petrol stations and from wood burning. The sulphur content in oil has also been controlled. Progress towards meeting the 2001 plan was reviewed in 2003.

The depletion of stratospheric ozone has been seen as a key environmental issue at the federal level in Canada. Monitoring of the ozone layer began in Canada in 1957 and from 1960 Canada became the home for the World Ozone Data Centre. Canada was the first country to sign the Vienna Convention for the Protection of the Ozone Layer (see Section 7.4.2) in 1986. The first protocol to this convention was signed in Montreal in 1987 and took the city's name. The use of some CFCs was banned in Canada before this international framework was in place. Three national plans have been published setting out how emissions of ozone depleting substances will be reduced. The latest of these is from 2001 (http://www.ec.gc.ca/ozone/en/index.cfm) and it also addresses the halocarbons brought in as substitutes for CFCs. Canada promoted an accelerated phasing out of CFCs and halons

TABLE 7.11 Canada NAAQOs acceptable ranges

Pollutant	Concentration	Measured as:
SO_2	334 ppb (900 µg m^{-3})	1-hour average
	115 ppb (300 µg m^{-3})	24-hour average
	23 ppb (60 µg m^{-3})	Annual average
NO_2	213 ppb (400 µg m^{-3})	1-hour average
	106 ppb (200 µg m^{-3})	24-hour average
	53 ppb (100 µg m^{-3})	Annual average
O_3	82 ppb (160 µg m^{-3})	1-hour average
	25 ppb (50 µg m^{-3})	24-hour average
	15 ppb (30 µg m^{-3})	Annual average
CO	30.6 ppm (35 mg m^{-3})	1-hour average
	13.1 ppm (15 mg m^{-3})	8-hour average
TSP	120 µg m^{-3}	24-hour average
	70 µg m^{-3}	Annual average

from 1998. By 2001, CFCs, methyl chloroform and halons were neither manufactured nor imported into Canada. The country's emissions of ozone depleting substances have fallen significantly.

In 1988, the Toronto Conference on the changing atmosphere was held and this focused attention on the climatic changes that might result from anthropogenic modification of the global atmosphere. Canada produced a National Action Strategy on Global Warming in 1990 in response to this. In 1992, Canada signed and ratified the UNFCCC. Canada has ratified the Kyoto Protocol. In 1994, a National Report on Climate Change was published, followed in 1995 by the National Action Program on Climate Change. In spite of its apparent enthusiasm to meet its international commitments, Canada has a high per capita level of energy use and high per capita CO_2 emissions. The most recent Climate Change Plan for Canada (Government of Canada, 2002) sets out a strategy to reduce Canada's annual emissions of greenhouse gases by 240 Mt (http://www.climatechange.gc.ca/english/index.shtml). The unwillingness of the USA to participate actively in the Kyoto process is identified as a significant problem: 'The decision by the Bush administration not to ratify the Kyoto Protocol poses an important challenge from both a climate change and a competitiveness perspective...For Canada, the US position presents unique challenges given our close economic relationship.' (Climate Change Plan for Canada, 2002, p. 7)

7.4 Global action

One of the most significant developments of the last 30 years has been the recognition that many air pollution issues require the development and implementation of policy at the supranational level. The issue of transboundary air pollution really came to the fore in the 1970s, following the recognition that environmental damage in Scandinavia was caused by emissions of acidifying pollutants in central and western Europe (see Chapter 4). The 1972 UN Conference on the Human Environment marked the start of the development of international cooperation in this area of policy. The emergence of issues which are clearly global in scale (the enhanced Greenhouse effect, stratospheric ozone depletion) has led to the development of policy structures to operate at this scale. The United Nations, through the Economic Commission for Europe (UNECE), the United Nations Environment Programme (UNEP) and the Framework Convention on Climate Change (UNFCCC), has been in the forefront of these developments.

7.4.1 UNECE Convention on Long Range Transboundary Air Pollution (LRTAP)

The UNECE was established in 1947 to promote economic recovery in Europe after World War II and now covers most of Europe, Canada, the USA and parts of central Asia. Although its initial remit was in terms of economic development, it was recognized early on that the organization could play an important role in setting safety and environmental standards. The UNECE Convention on Long Range Transboundary Air Pollution, signed in 1979, was the first internationally, legally binding, agreement to tackle air pollution across broad regions. Protocols to the convention have covered not only agreements to reduce emissions (Murley, 1995), but also set up a scientific and administrative framework to monitor and assess progress. There are currently eight protocols to this convention (Table 7.12).

The status of the protocols (signature, ratification) can be tracked through the UNECE website (http://www.unece.org/env/lrtap).

The 1984 Geneva Protocol, which set up EMEP (see Chapter 4), provided the foundations for the policy work. The protocol arranged the financing of a monitoring and modelling programme which collects emissions data, measures air and precipitation chemistry and models atmospheric dispersion (see Chapter 6). These provide the tools for assessing the present situation, developing new policy and assessing its likely impacts. There are a series of International Cooperative

TABLE 7.12 Protocols to the UNECE Convention on Long Range Transboundary Air Pollution

Year	Protocol	Popular name
1984	Long-term financing of the cooperative programme for monitoring and evaluation of the long-range transmission of air pollutants in Europe (EMEP)	Geneva Protocol
1985	Reduction of sulphur emissions and their transboundary fluxes by at least 30%	Helsinki Protocol ('The 30% Club')
1988	Concerning the control of nitrogen oxides and their transboundary fluxes	Sofia Protocol
1991	Concerning the control of emissions of volatile organic compounds and their transboundary fluxes	Geneva Protocol
1994	Further reduction of sulphur emissions	Oslo Protocol
1998	Heavy metals	Aarhus Protocol
1998	Persistent organic pollutants	Aarhus Protocol
1999	To abate acidification, eutrophication and ground level ozone	Gothenburg Protocol ('Multi-pollutant, multi-effect Protocol')

programmes addressing issues such as the effects of air pollution on forests, natural vegetation and materials, the development of critical loads (see Chapter 4) and a variety of modelling approaches. These programmes have their headquarters in different countries. The overall structure comprises a mix of intergovernmental bodies, expert groups and scientific centres. The political decision making is carried out at the highest level through the Executive Body (Fig. 7.1).

The initial approach within the UNECE was to adopt flat rate emissions reduction policies (e.g. the Helsinki Protocol or 30% Club). However, these did not recognize the spatial variability in ecosystem sensitivity or the patterns of pollutant transport across the UNECE area. It was realized that an effects-based approach, using the concepts of critical levels and critical loads (see Chapter 4), would be more effective. This came into play with the Sofia Protocol of 1988 and remains at the heart of policy making. The scope of protocols has also expanded, from dealing with single pollutants, to a range of pollutants with very different environmental and health effects. The latest

protocol, signed in Gothenburg in 1999, addresses three major environmental issues (acidification, eutrophication and ground level ozone) and sets emissions ceilings for European members of the UNECE for SO_2, NO_x, VOCs and NH_3. Canada and the USA have not agreed specific emission limits at this stage, but will set figures at ratification. There will be no targets for reducing NH_3 emissions for these two countries. The magnitude of emissions reductions agreed under the Gothenburg Protocol is considerable. Some examples are given in Table 7.13. Reductions are expressed as values for 2010 compared with the base year of 1990.

7.4.2 UNEP – The Vienna Convention on the Protection of the Ozone Layer

The United Nations Environment Programme started expressing concerns about the ozone layer in 1977. The Vienna Convention, signed in 1985, marked the first stage in the development of a global policy framework to protect stratospheric ozone. The Convention lacked any real substance, being largely a statement of

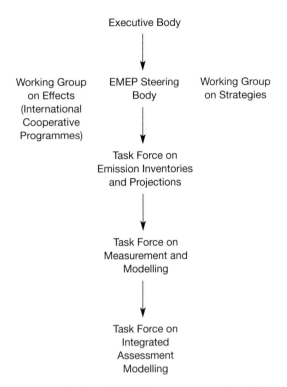

FIGURE 7.1 Institutional framework for the UNECE Convention on Long Range Transboundary Air Pollution

principle. It is, however, regarded as significant since it was the first attempt to address a global issue, one which at the time had few scientific underpinnings.

The recognition of severe depletion of ozone over the Antarctic in 1985 (see Chapter 3) prompted the signing of the Montreal Protocol on Substances that Deplete the Ozone Layer in

TABLE 7.13 Percentage reductions in emissions agreed under the Gothenburg Protocol relative to 1990 (UNECE, 1999)

Country	SO$_2$	NO$_x$	NH$_3$	VOCs
Czech Republic	−85%	−61%	−35%	−49%
Germany	−90%	−60%	−28%	−69%
Italy	−70%	−48%	−10%	−48%
Netherlands	−75%	−54%	−43%	−62%
Norway	−58%	−28%	0%	−37%
Spain	−65%	−24%	1%	−39%
Sweden	−44%	−56%	−7%	−54%
UK	−83%	−56%	−11%	−53%

1987. The 1987 protocol has undergone a series of adjustments and amendments as a result of meetings in London (1990), Copenhagen (1992), Vienna (1995), Montreal (1997) and Beijing (1999) (WMO, 1999). Details of these are available through http://www.unep.org/ozone/index.asp. The long-term goal of the protocol is to eliminate emissions of ozone depleting substances, but taking into account 'technical and economic considerations and bearing in mind the developmental needs of developing countries'. The protocol covers CFCs, halons, other halogenated CFCs, carbon tetrachloride, methyl chloroform, HCFCs, hydrobromofluorocarbons and methyl bromide. The consumption of CFCs, halons, halogenated CFCs, carbon tetrachloride and methyl chloroform is effectively banned in developed countries after 1996 (except for specified essential uses). Developing countries have until 2010 to meet these targets. Funds have been allocated to help developing countries with the development and adoption of more ozone friendly technologies. Ozone depletion is also being addressed through the Kyoto Protocol (see below).

7.4.3 UN Framework Convention on Climate Change (UNFCCC)

The Framework Convention on Climate Change was adopted at the UN Headquarters in New York in 1992 and opened for signature at the Earth Summit in Rio de Janeiro in the same year (http://unfccc.int/index.html). As with the Montreal Protocol, the long-term aim of the FCCC is ambitious: to stabilize emissions of greenhouse gases at safe levels. It is recognized that a combined strategy is needed, slowing climatic change and adapting to its effects. The scientific basis for understanding global climate change is provided largely by the IPCC, and the publication of the IPCC's first assessment report in 1990 (Houghton *et al.*, 1990) is seen as key to the development of the UNFCCC. The industrialized nations formed the Annex 1 parties to the Convention and had the aim (not legally binding) of keeping their emissions of greenhouse gases in 2000 to the same levels as in 1990, i.e. stabilization.

Developing countries formed a group called non-Annex 1 parties and their commitments were less specific.

However, it became clear that the FCCC was rather weak and, following discussions in Berlin (1995) and Kyoto (1997), the Kyoto Protocol was adopted in 1997. The Kyoto Protocol sets legally binding targets for greenhouse gas emissions for individual Annex 1 countries. The gases covered are CO_2, CH_4 (methane), N_2O (nitrous oxide), HFCs (hydrofluorocarbons), PFCs (perfluorocarbons) and SF_6 (sulphur hexafluoride). The target reductions for selected Annex 1 countries are shown in Table 7.14 and are to be achieved by the period 2008–12. It should be noted that not all countries are required to reduce their emissions.

A number of these compounds also play a role in stratospheric ozone depletion, hence the link between the Kyoto and Montreal Protocols. There are also strands relating to carbon sinks (forestry and land use). As with the Montreal Protocol, there is an emphasis on the transfer of technology to developing countries in the context of continuing, sustainable development. Emissions trading is also an important element (see below).

Although the Kyoto Protocol was signed by more than 80 countries, by late 2004 it can been ratified by 65, with Annex 1 country emissions at 44 per cent. It cannot come into force until it has been ratified by 55 countries, including Annex 1 countries responsible for at least 55 per cent of global CO_2 emission in 1990. Conferences of the parties (COPs) to the convention in Buenos Aires (1998) and Bonn (1999) have tried to clarify what is required. A meeting in The Hague in 2000 was notable only for its lack of success and resulted in there being a part 2 to COP6 which was held in Bonn in 2001. Although further COP meetings have been held every year (e.g. COP10 in Buenos Aires in 2004), the process was effectively stalled by the unwillingness of the USA to accept the terms of the protocol. Russia's decision means that the protocol will now come into force 90 days after official receipt of their ratification documents.

The members of the EU were committed to an overall emissions reduction of 8 per cent under the terms of the Kyoto Protocol. This meant that, under Article 4, separate targets

TABLE 7.14 Selected emission reduction targets for Annex 1 countries under the Kyoto Protocol (not all accepted by the countries identified)

Country (party to protocol)	Target (% change in emissions relative to base year/period)
Australia	+ 8%
Canada	–6%
Czech Republic	–8%
European Community	–8%
Japan	–6%
Poland	–6%
Russian Federation	0
USA	–7%

could be set for the member states depending upon their particular circumstances. These targets were agreed in 1998, ranging from –21 per cent for Germany to +27 per cent for Portugal. The UK's target was set at –12.5 per cent and it is expected that this will be met. The EEA report for 2003 (EEA, 2004) indicates that emissions of greenhouse gases across the EU have only fallen by 2.3 per cent between 1990 and 2001; this means that there has been very limited progress towards meeting the agreed Kyoto targets for 2008–12. Ten member countries are not likely to meet their national targets. Existing policy will probably only deliver a 0.2 per cent reduction in greenhouse gas emissions across the 15 EU countries by 2010 (compared with 1990). The main problem for Europe is emissions from transport (see http://www.themes.eea.eu.int/Environmental_issues/climate). The position of accession and candidate countries further complicates the picture of Europe's future greenhouse gas emissions.

In conclusion, it is clear that many countries are going to struggle to meet their Kyoto targets and that the targets themselves are not likely to be sufficient to prevent considerable climatic change (IPCC, 1995). The USA is, and has been, the world's largest emitter of greenhouse gases and its attitude towards Kyoto has served as a benchmark for other countries. At present, there are few grounds for optimism that significant reductions in greenhouse gases will be achieved.

7.5 Economic instruments

A range of economic measures, from taxes and emission charges to tradeable permits, can be used to encourage emissions reductions in addition to the more traditional approaches of specifying emission limits or requiring the adoption of particular pieces of technology (e.g. catalytic convertors for petrol vehicles). Increasingly, the use of economic incentives is seen as a way to meet environmental targets without damaging the potential for economic growth. This goal was set out in Principle 16 of the Rio Declaration on Environment and Development from the 1992 conference (http://www.unep.org/documents/default.asp?documentID=788&articleID=1163). Within the EU, for example, it has been suggested that the adoption of a community-wide emissions trading scheme could reduce the costs of meeting the EU's commitment under the Kyoto Protocol (see above), by 20 per cent (http://www.europa.eu.int/scadplus/leg/en/lvb/l28109.htm). This represents a significant sum of money.

The range and use of economic instruments has been reviewed by the EPA (2001 and 2002), but the focus here will be on the adoption and application of emissions trading through the buying and selling of allowances or credits. The idea that you could develop a market in emissions began in the 1960s (e.g. Crocker, 1966). The principle is that market forces will allocate emissions to the highest value uses, while those for whom emissions reductions can be achieved at lower cost, can sell their unneeded emission allocation on the open market. Emissions trading can take a number of different forms. According to the International Emissions Trading Association (http://www.ieta.org), these can be summarized as:

- Bubbles – where there is one entity for emissions accounting purposes which may comprise a number of individual sources (e.g. power stations operated by the same generating company).
- Offset or credit-based systems – where voluntary reductions or surplus emission capacity can be traded. Here the credit (i.e. the reduction in emission in excess of any required reduction) is the tradeable unit.
- Cap and trade programmes – where a regulating agency (e.g. the Environment Agency in England and Wales) sets a cap on emissions, usually based on a predetermined percentage reduction from historical emissions levels. If a regulated source (or sources) reduces its emissions below this cap, then the excess allowance can be traded. Under cap and trade programmes, the allowance is the tradeable unit of compliance. Any source must retain sufficient allowances to cover its own emissions over an accounting period. This form of emissions trading has been quite widely adopted and is discussed further below.

In order to ensure that you are not just moving emissions around through space, the size of the emission bubbles or caps will reduce through time.

The first emissions trading scheme in the USA was introduced in the 1970s for non-attainment areas under the original (1970) Clean Air Act (see above). Under this scheme, existing or potential new sources in non-attainment areas had to offset their emissions by buying emission reduction credits from existing sources (EPA, 2001). This ensured that there was no overall worsening of air quality. These arrangements were formalized through the 1977 amendments to the Clean Air Act. The most radical development, however, came with the 1990 Clean Air Act Amendments which

authorized a range of emissions trading programmes. Two of the best known of these are the Acid Rain Program and the Ozone Transport Commission (OTC) NO_x Program addressing the regional scale issues of acidification and low level ozone (see Chapter 4). The Acid Rain Program adopted a market-based approach to controlling SO_2 emissions from power stations, while taking a more traditional approach to NO_x control. The market in SO_2 began in 1995 under the first phase of the programme and is believed to have reduced US emissions by 4 million tons (ca. 4.06 tonnes). Unlike the nationwide Acid Rain Program, OTC NO_x focuses on 12 states in the north-east USA and Washington DC. Here there is a compliance period for NO_x sources from May to September with the goal of reducing high ozone concentrations. Another trading scheme aimed at controlling ozone levels was introduced by California's South Coast Air Quality District in 1993. Here, the Regional Clean Air Incentives Market (RECLAIM) covers NO_x and SO_2. The complexities of trying to limit the formation of ground level ozone are discussed further in Chapter 8.

The application of emissions trading which has attracted the greatest debate has been that relating to greenhouse gases through the Kyoto Protocol (see above). In this context, many countries are trying to agree an appropriate mix of emissions targets and how these might be achieved. Greenhouse gas emissions can be reduced by three routes: actual emissions cuts, the avoidance of potential emissions and the creation of additional sinks (e.g. carbon sequestration by planting new forests). For the Annex 1 countries (see above) these options are effectively expressed through three flexibility mechanisms: emissions trading, clean development mechanisms and joint implementation. Emissions trading and joint implementation apply only between Annex 1 countries, while clean development mechanisms involve private firms from Annex 1 countries meeting part of their emission reduction obligation by reducing emissions in developing countries. Perhaps unsurprisingly, this aspect has been controversial as it can be taken to mean that this will save some countries from any domestic emissions reductions at all. The development of emissions trading within the Kyoto framework is discussed in relation to Europe below.

For the purposes of Kyoto, the EU has an overall target, but retains flexibility for individual nation states (see Section 7.4.3). The UK has a target which is more ambitious than that for the EU as a whole and has moved ahead with policy formulation. The draft allocation of CO_2 equivalent emissions (http://www.defra.gov.uk/environment/climatechange/trading/index.htm) has been published and is consistent with a reduction in UK emissions of 16.3 per cent by 2010. Although the UK's main policy instrument is the Climate Change Levy (tax) introduced in 2001, there is also a voluntary emissions trading scheme for greenhouse gases (introduced in 2002). Under this scheme, emissions caps for some sectors (e.g. energy-intensive industries such as cement production) are determined by Climate Change Agreements. These agreements set specific targets to improve energy efficiency or reduce carbon emissions in return for getting a substantial reduction (80 per cent) in their liability to the Climate Change Levy. Here we see an example of a range of economic instruments coming together. The UK has developed its scheme ahead of that being put together by the EU (the EU Emissions Trading Scheme, EU ETS) which will come into force in 2005. The UK scheme is only a short-term measure (until 2007), when the EU scheme will take over as the primary trading mechanism. The UK is not the only EU country to have developed its own policy and there has been some debate over how well the national and EU schemes will co-exist (Boemare *et al.*, 2003).

As emphasized earlier in this chapter, the whole Kyoto process has been controversial and fraught with difficulty. The proposals for the use of economic instruments, including emissions trading, have been backed in some unexpected quarters (the USA), but rejected by others. In 2004, Australia walked out of the Kyoto emissions trading scheme on the grounds that there should be more emphasis on actual emissions reductions by the industrialized nations. The challenging issue of tackling global climate change is considered again in Chapter 8.

Successes, surprises and outstanding issues

8.0 Introduction

The preceding chapters have discussed a range of issues relating to atmospheric pollution and environmental change, ranging from monitoring and emissions (Chapter 2), through a series of major pollution concerns (Chapters 3, 4 and 5), modelling approaches (Chapter 6) and policy development (Chapter 7). While improved scientific understanding may be the primary goal of the scientific community, the value of atmospheric science to the wider world is usually gauged in terms of whether it has been able to make a useful contribution to better air quality and protection for the environment. The relationship between science and policy has become a key one (e.g. Sundqvist *et al.*, 2002), nicely illustrated by the fact that the IPCC produces summaries specifically for policy makers. In practice, science cannot address concerns ranging from local air quality to global climatic change without political will, the careful drafting of legislation and an effective system of regulation. Over the last 50 years, a number of air pollution issues have clearly been addressed with some success, in others the atmosphere has had some surprises up its sleeve and there are, undoubtedly, many outstanding issues. Some of these are discussed in the sections below.

8.1 Successes

Probably the most obvious success in restoring the atmosphere has been the sharp decline in smoke and SO_2 concentrations in most urban areas over the last 50 years. These classic urban air pollutants resulted in very poor air quality, low visibility, reduced sunshine hours and a range of environmental impacts in all cities where coal burning was the dominant fuel type, particularly for domestic use. From early beginnings, this 'original' smog came to dominate urban areas through the industrial revolution. As smoke control legislation and changes in domestic fuel use improved conditions in the developed world, so major cities in the developing world came to experience the same poor conditions. Early pollution in cities such as Mexico and São Paulo was of this type. In Asia, conditions in cities such as Beijing came to match the conditions that had prevailed in Europe in the early to mid-twentieth century. Even in these cities of the developing world, however, controls on large industrial sources of SO_2 and a shift from coal to gas for domestic use have lowered levels of these 'traditional' pollutants (see Chapter 5). The poor visible air quality prevalent in most cities until the 1960s is unthinkable today. In practice, the issue was a straightforward one and alternative, cheap technologies were readily available.

A second issue that can be considered to have been tackled with some degree of success is 'acid rain'. By the late 1960s, industrial and economic development in Europe had become reliant on coal and oil to generate cheap electricity in large power stations. As described in Chapter 4, scientists in Sweden began to notice the deterioration in the fish stocks in their lakes and attributed this to the acidification of both the lakes and the stream-waters that fed them. The finger of suspicion pointed to the large electricity-generating power stations with their high stacks which acted as huge point sources of sulphur dioxide, the major precursor to the damaging acidic species. The decline in fishery status of remote lakes and streams, the acidification of soils and concerns about forest decline spread from Sweden to much of Scandinavia and elsewhere in Europe, and even to North America. The original attribution of blame has proved to be largely correct, namely the SO_2 from power stations, though NO_x has been added into the equation.

The environmental impacts of acidification have been substantiated and the European nations have moved to combat acidification by cooperative action through the United Nations Economic Commission for Europe and the Commission of the European Communities (see Chapter 7). During the 1990s, SO_2 emissions declined dramatically, with each nation taking a different approach to mitigate emissions. Germany has taken the best available technology approach and fitted flue-gas desulphurisation to their large coal fired power stations. In contrast, the United Kingdom has switched a significant fraction of its electricity generation capacity to natural gas fired combined cycle turbine plant with a concomitant reduction in CO_2 emissions. Such steps, together with a general reduction of the sulphur content of petroleum fuels, have secured a halving in European SO_2 emissions and reduced acid deposition loads. The chemistry of oxidized sulphur is relatively straightforward and large, point sources of SO_2 easily identified and controllable given appropriate investment. The shift of attention to nitrogen deposition as a potential source of acidity has raised new issues in terms of the nature of the sources and the availability/practicality of emissions controls. Questions about vehicle numbers and types are enmeshed with lifestyle choices (e.g. the rise of the sports utility vehicle in the USA has adversely affected NO_x emissions) and there is no easy technological fix for emissions from livestock as there might be for a factory chimney. Nitrogen also has a much more complex chemistry both in the atmosphere and in its interactions with the biosphere.

Given that policies to reduce emissions of SO_2 have been in place in Europe for 20 years, it is reasonable to ask whether these have 'delivered' the reductions in S deposition that might have been hoped for. Reductions in emissions of oxides of N have been more recent and smaller, but the same question can be asked. In the UK context, the National Expert Group on Transboundary Air Pollution (NEGTAP, 2001) was asked to consider how the concentrations and depositions of pollutants had changed in response to international protocols and whether ecosystems were, or would be, protected. Analysis of data from the UK's monitoring networks (see Chapter 2) has shown that although the overall decline in S deposition over the period 1986–97 is roughly equal to the reduction in emissions, this decline has not been equally split between wet and dry deposition. There has also been a clear pattern to changes across the country. Dry deposition has decreased more rapidly than wet; most of the improvements have been in central and eastern England (i.e. close to the major source areas where SO_2 concentrations have fallen quite dramatically) and least in upland, high rainfall areas in the west of Britain (i.e. remote from source areas). The content of non-sea salt sulphate has declined across the UK, but in remote western sites this trend is only statistically significant if all the sites are grouped together (NEGTAP, 2001). In the English East Midlands, S deposition has declined by more than the national reduction in emissions. Trends in nitrate and ammonium in rain have been smaller than those for sulphate, but show a similar 'geography' to the latter. Only nitrate close to source areas showed a statistically significant downward trend. Similar responses to emissions reductions have been reported for the USA by Civerolo et al. (2001).

A number of processes may be involved in the apparent 'non-linearity' between sulphur emissions and deposition, including: different patterns of emission reduction between high and low level sources, changes in the rates of dry deposition of SO_2 (affected by NH_3), changes in oxidation pathways and changes in background concentrations of non-sea salt sulphate. It is apparent that both the chemical reactions taking place in the atmosphere and rates of removal have changed through time as the chemical composition of the atmosphere has been changing. Maintaining an understanding of these changes is important for both atmospheric scientists and policy makers.

8.2 Surprises in atmospheric pollution

Over the last 50 years, the industrialized countries have become concerned by a series of environmental surprises involving atmospheric pollution. Since these surprises were by their very nature unpredicted, they have acted as a wake-up call to policy makers and the atmospheric science community.

The rapid industrialization of southern California during the 1940s led to the occurrence of a choking and irritating smog accompanied by a visibility-reducing haze which has become known as Los Angeles smog (see Chapter 5). Early attempts by policy makers to control this new form of air pollution focused on smoke and sulphur dioxide, the old well-understood pollutants, and so they were doomed to failure as they were not responsible for this new form of smog. A start was made to understand the atmospheric processes involved in photochemical smog formation and to characterize the damaging and causative agents involved (Haagen-Smit *et al.*, 1953). These early studies broke new ground and led to the first air pollution inventories and models, the first application of air quality management and the first laboratory investigations of complex environmental problems.

Understanding the mechanism of Los Angeles smog formation took decades of work

combining atmospheric monitoring, laboratory investigations, field studies and computer modelling. Policy makers' efforts were focused on controlling emissions of organic compounds since these were found to be the key precursors to the formation of ozone and other oxidants in the presence of oxides of nitrogen and sunlight. As described in Chapter 5, peak levels of ozone have now begun to fall and the intensity and spatial extent of the exceedances of the ozone air quality standard have steadily decreased. All this improvement has been achieved despite half a century of unparalleled growth in the industrial and economic development of southern California.

Although photochemical smog was first identified in southern California, it has subsequently been found elsewhere in the USA with some subtly different characteristics. Photochemical smog is, or course, not confined to the USA, being a major issue in places as diverse as Athens and Santiago. It is also now the dominant concern in Mexico City (see Chapter 5). However, as photochemical smog first emerged in the USA, its diversity is best illustrated from that country.

Elevated ozone concentrations have been found along the eastern seaboard of the USA where their occurrence has been seen as a regional phenomenon covering cities from Philadelphia, Baltimore, Washington DC, New York and Boston (up to Maine), rather than as the purely urban phenomenon of the Los Angeles Basin (Wolff *et al.*, 1977). Policy makers have been surprised how the elevated ozone levels have been stubbornly resistant to controls on organic compound emissions and they have now switched to policies based on controls on both organic compounds and oxides of nitrogen. This reflects the importance of understanding whether ozone production is VOC- or NO_x-limited, as described in Chapter 4. Reductions of the appropriate precursor are the only way to reduce ozone concentrations; understanding the atmospheric chemistry is the key to successful policy.

The ozone air quality problem in the USA has exhibited one further surprise beyond those in southern California and the north-eastern seaboard. Houston, Texas has taken over the lead for the highest number of exceedances of the ozone air quality standard during the late 1990s. The new ozone problems were subtly different from those experienced elsewhere, and again took policy makers by surprise. Ozone air quality in Houston is dominated by intense plumes of ozone that waft over populated areas. The source of these intense plumes involves emissions of reactive olefins from specific petrochemical installations. Understanding Houston's air quality problems has required the focusing of inventories, field studies and modelling on to the handful of key reactive olefin species, and improving air quality will require their efficient emission controls (Ryerson *et al.*, 2003).

While air quality policies to combat photochemical smog have met with some success, this is by no means the case for all environmental surprises that have become apparent during the second half of the twentieth century. Where the surprises have involved processes with strong positive feedbacks or where the main culprits have long atmospheric lifetimes, there are clear limits to the success of the policy actions to combat them. In this sense, the genie is out of the bottle. Pandora's box has been opened.

An example of an environmental surprise that has been difficult to combat, despite good progress with policy, is the depletion of the stratospheric ozone layer by chlorofluorocarbons (CFCs) (see Chapter 3). This is the story of the widespread use of a set of chemicals which had unimagined environmental consequences. There is no question that the CFCs have been extremely useful chemicals because they are inert, stable, harmless, non-toxic and straightforward to manufacture. Their application began with refrigeration during the 1930s, harnessing their unique thermodynamic properties, which made them ideal working fluids for domestic and commercial appliances. Their use as aerosol propellants first arose during World War II to provide easy to use insecticides for military operations within the tropics. Their use as foam-blowing agents and in air-conditioning equipment began during the 1960s.

By the end of the 1960s, global CFC production had reached several hundreds of thousand tonnes per year. Jim Lovelock, the inventor of the most sensitive analytical detector, the electron capture detector, showed that the CFCs were accumulating steadily in the atmosphere and that almost all the CFCs ever released to the atmosphere were still there (Lovelock, 1972). The CFCs were inert and stable at least in the troposphere. Molina and Rowland (1974) hypothesized that the CFCs would eventually reach the upper atmosphere and be transported above the ozone layer where they would be photolysed by solar ultraviolet radiation, decomposing into reactive chlorine compounds. These reactive chlorine species would catalyse the depletion of the ozone layer and allow damaging ultraviolet radiation to reach the ground. In 1985, Joe Farman published his observations of the spring-time decline in the total ozone burden above Halley Bay in Antarctica over the nearly 30-year period since his measurements had begun in 1957 (Farman *et al.*, 1985). The concept of the 'ozone hole' was born. Within a few years, airborne scientific missions to the ozone hole and satellite observations confirmed its occurrence and confirmed that the photolysis of the CFCs and the degradation of other ozone depleting substances were ultimately responsible.

Although the Rowland and Molina hypothesis came as a surprise, the response of the policy makers in North America and Europe was rapid. A freeze was agreed on the building of new CFC production capacity and all non-essential uses in aerosol spray cans were phased out. Negotiations began on an international convention to protect the stratospheric ozone layer and this was signed during 1985 (see Chapter 7). Long-term monitoring was established for the CFCs and industry began the search for alternatives. As a result, atmospheric emissions of CFCs stabilized but atmospheric concentrations continued to rise. The Montreal Protocol (1987) accepted the distinction between essential and non-essential uses of the CFCs and introduced the concept of the ozone depleting potential (ODP) as a means of indicating the relative importance of the different ozone depleting chemicals. Phase-out schedules were agreed for each ozone depleting chemical based on its ODP. Due recognition was given to those countries with economies in transition and to those with development problems.

The Montreal Protocol was a brave and ambitious start to global negotiation. However, by the early 1990s, satellite observation was showing how year after year the ozone hole over Antarctica was growing in depth and land area and persisting longer into the summer. The atmospheric concentrations of the CFCs were continuing to grow. In 1990, policy makers agreed to the London Adjustments and Amendments and subsequent adjustments and amendments were agreed at three-year intervals (see Chapter 7). These amendments steadily brought forward the phase-out dates for production and sales of the ozone depleting chemicals and questioned all claims that uses were essential. Industry responded with a range of hydrochlorofluorocarbon substitutes which had lower ODPs because they carried less chlorine per molecule and had dramatically shorter atmospheric lifetimes because they were degraded in the lower atmosphere.

By the early 2000s, HCFC substitution had replaced all essential uses and they were themselves beginning to be phased out as well, in favour of hydrofluorocarbons (HFCs) which contain no chlorine whatsoever. Although atmospheric levels of methyl chloroform have declined dramatically because of its five-year atmospheric lifetime, most CFC concentrations have shown little more than a stabilization. Indeed as of 2003, atmospheric concentrations of CFC-12 continue to rise, because of its long lifetime (100 years) and its continued emission from old refrigerators and foams. It will be only on the century timescale that atmospheric burdens of the CFCs will begin to fall dramatically and with them the reactive chlorine content of the stratosphere (Fig. 8.1). Ozone holes will be an annual occurrence for much of the twenty-first century and it will be decades before the policy makers can claim that stratospheric ozone depletion has been controlled.

The issue of climatic change resulting from human actions falls into a similar category to ozone depletion. It is easy to forget that up until the late 1970s, the concern was that we might be slipping into the next ice age. Forecasts of possible nuclear winter only reinforced the impression that the global climate was likely to get significantly colder, not significantly warmer.

There are many difficulties in grasping the issue of climate change: its global scale, the long lived nature of many of the pollutants concerned and its complexity. Many pollutants otherwise implicated in acidification or stratospheric ozone depletion also play a role in radiative forcing (see Chapter 3). Controls on sources of anthropogenic aerosols and success in preventing stratospheric ozone depletion will have the unfortunate side effect of reducing negative forcing. A number of the compounds brought in to replace CFCs are highly efficient greenhouse gases increasing the positive radiative forcing in the atmosphere.

Concerns over climatic change caused by human actions have increased interest in the natural variability of the global climate system. For many people, this has also been a surprise. We now know that the global climate is able to change significantly over very short timescales (decades or less). We are likely to get little or no warning that one of these shifts is underway and it is still unclear how far anthropogenic drivers will interact with features of natural variability such as ENSO, the North Atlantic Oscillation and the Pacific Decadal Oscillation (IPCC, 2001). Even more worrying for those in north-west Europe is whether anthropogenic warming might stop deep water formation in the North Atlantic, effectively shutting down the conveyor system that brings warm waters into high latitudes in this part of the Northern Hemisphere. Getting dramatically colder as the rest of the world warms might be considered to be the ultimate surprise.

8.3 Matters of scale

The traditional approach of much work in the atmospheric sciences has been to focus on an issue and set of pollutants which appear to be important at a particular spatial scale. Aspects of global (Chapter 3), regional (Chapter 4) and urban (Chapter 5) air quality have generally been studied in isolation from each other. In the case of pollutants with short atmospheric lifetimes, there may be some justification for this, but for many, such divisions are entirely artificial. NO$_x$, which might cause local air quality problems at kerbside in an urban area, may go on to generate NO$_3$ with regional scale effects through acidification and eutrophication. SO$_2$ might affect local lichen populations, be a source of regional scale acidification and play a

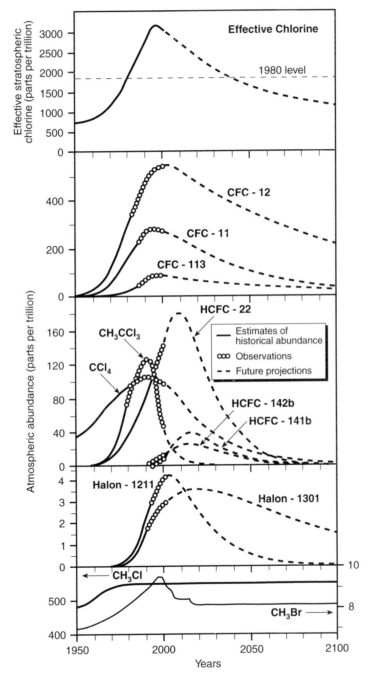

FIGURE 8.1 Past and future abundance of atmospheric halogen source gases (Source: WMO, 2002 (Fig. q16–1))

role in the global radiation balance when in its aerosol sulphate stage. Ideally, we should think of a continuum through both space and time (Fig. 8.2). The relationship between science and policy making structures has meant that there has often been limited consideration of the effects of policy developed at one scale for issues largely considered at a different scale. Over recent years, there have been moves to integrate, in an explicit manner, issues such as air quality, acidification and eutrophication. This has been reflected at the national level (e.g. in the Netherlands) and at the international level. A strand of this is the shift from policies addressing one pollutant at a time to a multi-pollutant, multi-effect approach. The Gothenberg Protocol to the UNECE

Convention on Long Range Transboundary Air Pollution (see Chapter 7) typifies the latter approach.

Another important aspect of spatial scale has come with the recognition that achieving local air quality targets for some pollutants may be profoundly affected by changes in background concentrations at the regional, or even hemispheric scale. As described in Chapters 4 and 6, this is certainly the case for tropospheric ozone. Peaks in ozone concentrations have been reduced across much of Europe and North America, but rising background concentrations (Simmonds *et al.*, 2004) pose a threat to our ability to meet targets set to protect both human health, crops and natural ecosystems from longer-term exposures. Both measurement

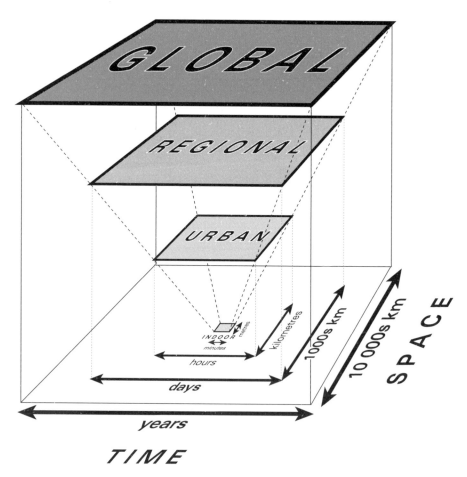

FIGURE 8.2 Spatial and temporal continuum of atmospheric pollution issues

campaigns and modelling studies have indicated ultra long-range (i.e. intercontinental) transport of low level ozone. Modelling can also be used to quantify the contribution from different sources and the spatial variability of their influence. A study by Derwent *et al.* (2004) has indicated that, as an annual average, about 7ppb of ozone across Europe originates in North America and about 5ppb in Asia. The distribution of these intercontinental contributions varied both spatially and temporally across Europe, indicating that emissions controls within Europe alone would have differential impacts in different areas.

Modelling approaches have also tended to follow these spatial categories (local, regional, global) (see Chapter 6). This can partly be explained by available computer power; to run a global model at the spatial and temporal resolution required to address local air quality issues would not be practical. Much of the input data would not be available at the appropriate resolution and the run times would be unsupportable. The desirability of integrating modelling systems at different scales, perhaps by nesting models one inside the other, is widely recognized. To some extent the US EPA has addressed this through using a family of models under Models-3 (Chapter 6). Within Europe, EMEP are developing a system of nested models for photochemical pollutants and particles (e.g. Wind *et al.*, 2002).

Perhaps one of the most surprising divides has been that which seems to separate those interested in atmospheric pollution from those interested in climatic change. Some of this may again relate to spatial scale, but may also reflect such issues as disciplinary background (largely chemists vs physicists). Climatic change also brings with it a very different sense of timescale ('deep' or geological time) from conventional air pollution concerns. The involvement of earth scientists and historians has brought a whole new perspective to the study of the earth's climatic history. Some of their contribution through the study of cores from lakes, ice sheets and oceans, has been outlined in this book. The integration of climatic change and traditional regional air pollution issues is now starting to happen within the scientific community. In Europe this linkage is represented by the AIR-CLIM Project which seeks to adopt some of the methodologies used to study regional air pollution to explore the effects of climate change and to bring these two strands together (e.g. Alcamo *et al.*, 2002).

Divisions, however, persist in many policy making organizations such as the Department for Environment, Food and Rural Affairs (DEFRA) and the Environment Agency (EA) in the United Kingdom. Such divisions mean that there have been few efforts to develop abatement strategies which seek to address both atmospheric pollution and climatic change issues simultaneously. It is clear, however, that policies to reduce emissions of greenhouse gases (especially CO_2) are likely to improve a range of other air pollution issues and to reduce the costs of separate policies to reduce pollutants such as SO_2, and to a lesser extent, NO_x. The health benefits that might be associated with reducing emissions of greenhouse gases (separate from the benefits of restricting temperature increases), have been quantified for four major cities by Cifuentes *et al.* (2001). These authors calculated that adopting existing technologies to reduce greenhouse gas emissions would have the effect of reducing particulate matter and ozone concentrations by 10 per cent. With respect to their four study areas (Mexico City, New York, Sao Paulo and Santiago), they indicate that this reduction in pollutant load would avoid 64,000 premature deaths, 65,000 cases of chronic bronchitis and 37 million person days of restricted activity or loss of work. They regard these estimates as being conservative.

8.4 Outstanding issues

As indicated in Chapter 7, perhaps the outstanding issue in relation to the global atmosphere is whether humankind will show the will to control emissions of gases implicated in global warming. The scientific case for anthropogenically induced global change appears to get stronger and stronger. The IPCC Third Assessment (IPCC, 2001) stated that 'In the light of new evidence and taking into account the remaining uncertainties, most of the observed warming over the last 50 years is likely to have been due to the increase in greenhouse gas concentrations'. They also make it clear that even if there were the political will to reduce emissions immediately, anthropogenic climate change will persist for many centuries. There are, of course, dissenting voices opposing this mainstream view. Perhaps the best known of these is Lomborg (2001); although generating much discussion, his views have had little impact on the scientific mainstream.

There are those that contest the science, but an important issue for some is that of uncertainty. The global climate system is hugely complex, and almost every aspect has uncertainties associated with it: instrumental and proxy records of change, present and future estimates of sources and sinks of key trace gases and aerosols, the internal natural variability of the climate system and, of course, the computer models used to test the behaviour of the climate system in response to different forcings. A number of these uncertainties can be quantified and presented so that both the general public and politicians can see their effects. Estimates of global temperature change and sea level rise under the range of IPCC scenarios while trying to account for uncertainties, are presented in Fig. 8.3. The absolute values may vary, but the trends are consistent.

Gupta et al. (2003) have explored the problem of uncertainty in emissions inventories and in assessing compliance to reductions agreed through the Framework Convention on Climate Change (FCCC) (the Kyoto Protocol, see Chapter 7). They suggest that the uncertainties in the emissions inventories are as large as the targets for emissions reductions. In particular, they highlight the problems caused by the adoption within the FCCC of a 'basket' of gases approach rather than a gas-by-gas basis. This allows governments to adopt a range of approaches to reducing emissions of a range of gases without the requirement to quantify the uncertainties in meeting targets inherent in these different combinations. The authors suggest that there are a number of dangers in the current approach which might allow governments to go for options which apparently meet commitments in the short term, minimizing socio-economic (lifestyle) impacts, but which fail to address the long-term goals of the FCCC.

There is no doubt that scientific uncertainty in issues relating to all aspects of atmospheric pollution can make uncomfortable reading for policy makers. Uncertainties are often large; it has been suggested that any air pollution model which achieves results within a factor of two of a measured value, is doing well. This may not conform with the general public's view of good model performance. Quantifying and understanding uncertainty is significant in terms of both initial policy development and in determining whether an agreed policy will actually achieve its environmental protection goals. Colvile et al. (2002) have looked at the issue of uncertainty in the context of local authorities in the UK and the declaration of Air Quality Management Areas (AQMAs) under the Air Quality Strategy (see Chapter 7). Some authorities may wish to adopt a precautionary approach and declare AQMAs large enough to include all possible areas where targets might be exceeded, while others may wish to avoid declaring AQMAs, due to possible negative effects such as falling property values and restrictions on vehicle use. Although uncertainties from air quality models can be quantified, many of the political and social issues in declaring an AQMA (or not) cannot be quantified so readily. In many areas of atmospheric pollution there seems to be growing enthusiasm for the use of probabilistic approaches, including mapping. Such an approach can provide a transparent and easily visualized means of conveying the effects of scientific uncertainty to those trying to develop effective environmental policies.

While the success, or otherwise, of policy is most readily assessed through the measurement of changing atmospheric concentrations and depositions of the compounds of interest, if the long-term goals of environmental protection are to be met, then ideally there should be biological recovery as well as chemical recovery (e.g. in freshwater ecosystems). The difficulties of identifying biological recovery are, however, considerable. Because surface waters might be expected to respond relatively quickly to changes in atmospheric inputs, much focus has been on whether acidified lakes and streams have shown a biological response to the substantial decrease in emissions of SO_2 (and to a lesser extent NO_x) over the last 20 years or so. Stoddard *et al.* (1999) looked at water chemistry data from 205 sites in Europe and North America and found generally strong declines in SO_4 concentrations, but little (if any)

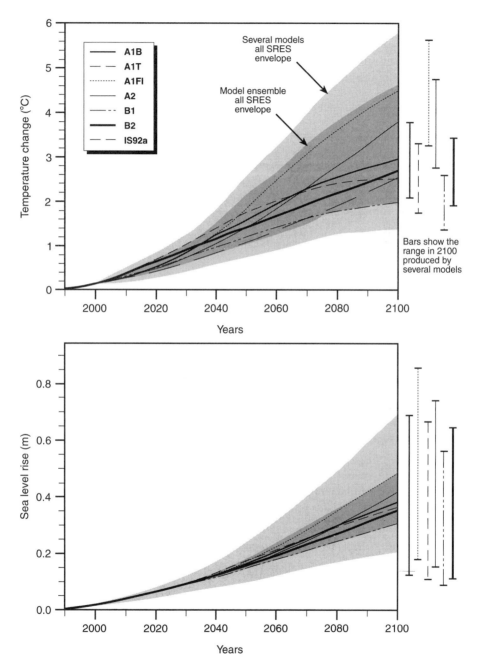

FIGURE 8.3 Estimates of (a) global temperature change and (b) global sea level rise under a range of IPCC scenarios showing the uncertainty ranges of those estimates (Source: IPCC, 2001)

decline in NO_3. Surface water alkalinity was found to have recovered in many sites. The chemical response was found to be quite highly variable spatially, something also reported in later studies. They did not address the issue of biological recovery but, at the time, there was no clear evidence for it.

More recent reviews of recovery from acidification have been published by Jenkins *et al.* (2001) and as a special issue of *Ambio* (2003). Data from severely acid damaged lakes around Sudbury, Ontario have shown evidence for biological recovery in algae, macroinvertebrates and fish. What has become clear, however, is that climate change and the introduction of exotic species (primarily fish) is altering the species composition of high latitude systems and effectively changing their potential biological recovery end point. In the UK context, NEGTAP (2001) found clear evidence for chemical recovery but little evidence of biological recovery. Time lags between changes in emissions and changes in water and soil chemistry, species dispersal and recolonization, the impacts of pollution episodes and climatic variability will all affect the nature and rate of biological response. The identification of a desirable target biological status is a separate and complex issue.

8.5 Do we have the tools to meet the challenges?

It is undoubtedly the case that the level of understanding of the chemistry and behaviour of the atmosphere has increased significantly over the last 20 years. The linkages between atmospheric composition and climate also become ever more apparent. Questions have been asked, however, about how well placed we are to identify, understand and address issues of atmospheric change over the longer term. A serious failing seems to be the lack of long-term observations (National Research Council (NRC), 2001). Most monitoring networks are set up to meet specific objectives and have relatively short lifetimes (see Chapter 2). Sites can usually only monitor a limited range

of compounds and may have no meteorological data capture. Co-location of sites in different networks is quite rare because of policy imperatives and cost.

The NRC (2001) report suggests that current observational networks are not adequate to address many important global air quality changes. Particular weaknesses are identified as being: long-term measurements of reactive compounds and particulate matter, a lack of vertical profile data and a lack of sites which allow an assessment of long-range transport and background concentrations. They also point out the need for global standards for measurements and to integrate measurements from different programmes and platforms (i.e. satellite, aircraft and ground-based data). It is also clear, at a range of scales, that measurement data are often ill-suited to the development and validation of models. Poor spatial distribution, a lack of multiple parameter measurements at the same site and short periods of data collection at sites all mitigate against good quality model testing and validation. The effectiveness of policy can only be assessed in the longer term if relevant monitoring programmes are maintained. The NEGTAP (2001) report, which attempted to assess changes in pollution exposure and ecosystem response over the last 20 years in the UK, highlighted how difficult this long-term approach proves to be in practice.

Impact assessment (e.g. actual acidic deposition in a catchment, or exposure to pollution of a set of individuals) is another significant issue. In terms of human health, personal exposure to pollutants, especially indoors, remains poorly quantified; the role of long-term exposure to pollution and the effects of combinations of pollutants remain poorly understood. For ecosystems, the effects of air pollutants have to be considered in the context of climatic variability, natural predators, insect and fungal infestations and natural population dynamics. In the case of natural ecosystems, a complex history of pollutant exposure and climatic history may have to be taken into account. In managed systems (including agricultural crops), changing management practices and market priorities will also have to be considered. Monitoring these impacts is a complex process which requires the long-term commitment of substantial funds.

8.6 Conclusions

No single book can hope to address the huge range of issues covered by atmospheric pollution. Here the focus has been on the best known pollutants that affect both the natural and human environment. For the sake of convenience, pollutants have been considered at the scale where they are usually felt to be of the greatest concern (global, regional, urban), but it has been emphasized that any such division is artificial and should not stand in the way of taking the broader view and considering how the composition of the atmosphere varies on all scales, both spatial and temporal. Most direct measurements of the atmosphere only cover short periods of time, but there are methods which allow long-term changes (centuries, millennia) to be assessed. Such long-term understanding is key if we are to understand how we reached our current situation and where we might be going in the future.

This book has only touched on some issues (e.g. heavy metals) and said little or nothing about others which are known to be of considerable interest (e.g. organic pollutants, the role of biomass burning, the impact of aircraft emissions). Atmospheric and climatic sciences are a fast developing field and this textbook is only intended to provide a few foundations. The interested reader is advised to turn to the wider literature. Traditional boundaries between disciplines, or fields of study, are breaking down. While science can underpin policy making, meeting the challenges of air pollution and atmospheric change ultimately depends on society's priorities and willingness to act.

GLOSSARY

ADORC Acid Deposition and Oxidant Research Center (Japan).

Aerosols The dispersion of solid or liquid particles in the atmosphere, the latter remaining airborne for at least several hours.

AGAGE Advanced Global Atmospheric Gases Experiment, a monitoring programme for ozone depleting substances and greenhouse gases.

AGCM Atmospheric General or global Circulation Model.

AGMAAPE Advisory Group on the Medical Aspects of Air Pollution Episodes (United Kingdom).

AIR-CLIM An activity linking Air Quality and Climate Change.

AIRBASE This is the air quality information system of the European Environment Agency.

AIRMoN This is the Atmospheric Integrated Monitoring Network of the National Oceanic and Atmospheric Administration in the United States of America.

ALE Atmospheric Lifetime Experiment, a monitoring programme for ozone depleting substances.

AMIS Air Management Information System of the World Health Organization.

ANC Acid Neutralizing Capacity.

AOGCM Atmosphere-Ocean General or global Circulation Model.

AOP Auto-Oil Programme, a joint activity between the motor manufacturers and the oil industry in Europe to consider future air quality issues.

AOT$_{40}$ Accumulated Ozone above a Threshold of 40 ppb is an ozone exposure index designed for studies of the impacts of ozone concentrations on crops and vegetation systems.

AOT$_{60}$ Accumulated Ozone above a Threshold of 60 ppb is an ozone exposure index designed for studies of the impacts of ozone concentrations on human health.

AQMA Air Quality Management Area is a pollution control term used in the United Kingdom.

ARIC Acid Rain Information Centre (Manchester Metropolitan University, United Kingdom).

ATMES Atmospheric Transport Model Evaluation Study.

B[α]P Benzo[α]pyrene is a polycyclic aromatic compound.

BACT Best Available Control Technology, a pollution control term used in the United States of America.

BAPMoN Background Air Pollution Monitoring Network.

BART Best Available Retrofit Technology, a pollution control term used in the United States of America.

BAS British Antarctic Survey (United Kingdom).

BATNEEC Best Available Technology Not Entailing Excessive Costs, a term used in pollution control (United Kingdom).

BPEO Best Practicable Environmental Option, a term used in pollution control (United Kingdom).

BPM Best Practical Means, a term used in pollution control (United Kingdom).

CACGP Commission on Atmospheric Chemistry and Global Pollution.

CAPMoN Canadian Air and Precipitation Monitoring Network (Canada).

CARB California Air Resources Board (California, United States of America).

CASTNET Clean Air Status and Trends Network (United States of America).

CBr$_y$ The total organic bromine content of the atmosphere obtained by summing the concentrations of the different organic bromine carriers or bromocarbons.

CCC Chemical Coordinating Centre, a scientific centre within the EMEP programme.

CCE Coordinating Centre for Effects, a scientific centre within the EMEP programme hosted at the RIVM.

CDIAC Carbon Dioxide Information Analysis Center of the Argonne National Laboratory, USA.

CFCs Chlorofluorocarbons, compounds whose molecular structures contain carbon, chlorine and fluorine that take part in stratospheric ozone depletion.

CFD Computational Fluid Dynamics, the study of fluid flow.

CL A Critical Load is a quantitative estimate of exposure to one or more pollutants below which significant harmful effects on specified sensitive elements of the environment do not occur according to present knowledge.

CMAS Community Modeling and Analysis System (United States of America).

COMEAP Committee on the Medical Effects of Air Pollutants (United Kingdom).

CORINAIR Core Inventory of Air Emissions in Europe, an activity within the scope of the European Environment Agency, Copenhagen, Denmark.

CRI Chemical Release Inventory of the Environment Agency (United Kingdom).

CTM Chemistry Transport Model.

CWS Canada Wide Standards (Canada).

DEFRA Department for Environment, Food and Rural Affairs (United Kingdom).

DETR Department of the Environment, Transport and the Regions (United Kingdom).

DF Distrito Federal (Mexico).

DIRTMAP Dust Indicators and Records of Terrestrial Marine Palaeoenvironments.

DMS Dimethyl sulphide $(CH_3)_2S$, a natural constituent of the atmosphere derived from biological processes occuring in the surface layers of the oceans.

DU Dobson Unit, is the thickness that the ozone in the atmosphere would have at that point if it was held at standard temperature and pressure, and is recorded in milli-atmo-centimetres. 1 DU is 2.687×10^{16} molecules per cm^2

EA Environment Agency (United Kingdom).

EACN European Air Chemistry Network, a monitoring programme operated by the Department of Meteorology, University of Stockholm, Sweden.

EANET This is the Acid Deposition Monitoring Network in East Asia.

EC European Community.

ECN Environmental Change Network (United Kingdom).

EDGAR Emission Database for Global Atmospheric Research with information on emissions by region and by source for greenhouse gases, ozone and aerosol precursors.

EDGAR-HYDE This is an Emission Database for Global Atmospheric Research for anthropogenic emissions over the period from 1890–1990.

EEA European Environment Agency.

EEC European Economic Community.

EFTA European Free Trade Area.

EKMA-OZIPM Empirical Kinetic Modelling Approach Ozone Isopleth Model.

EMEP European Monitoring and Evaluation Programme, a programme of the United Nations Economic Commission for Europe and its Convention on Long Range Transboundary Air Pollution.

Emission inventory This is a tabulation of the masses of a range of pollutants discharged into the atmosphere from a given process, in a given length of time, in a given location.

ENSO El Niño Southern Oscillation is a coupled atmosphere-ocean phenomenon in which a pool of warm water develops off the coast of Ecuador and Peru, associated with a fluctuation in the intertropical surface pressure pattern and circulation of the Indian and Pacific Oceans. It has significant effects both locally (fisheries) and on the global climate.

EPAQS Expert Panel on Air Quality Standards (United Kingdom).

ESA European Space Agency.

ETC/ACC The European Topic Centre on Air and Climate Change is an activity of the European Environment Agency.

ETEX European Tracer Experiment.

EU European Union.

EUROSTAT This is the Statistical Office of the European Communities.

EUROTRAC The European Experiment on the Transport of Environmentally Relevant Trace Constituents over Europe, a programme within the EUREKA initiative of the European Union.

GAGE Global Atmospheric Gases Experiment, a monitoring programme for ozone depleting substances.

GAW Global Atmosphere Watch, a programme of the World Meteorological Organization.

GCM General or global Circulation Model.

GCOS Global Climate Observing System, a programme of the World Metoeorological Organization.

GEIA Global Emissions Inventory Activity, a programme of IGAC.

GEMS/AIR A global urban air quality management programme organized by the WHO and UNEP.

GIS Geographical Information System.

GISP$_2$ The US Greenland Ice-Sheet Project (United States of America).

Global scale This refers to the spatial scale of a process or a model that is of the order of 10,000 kilometres or longer.

GO3OS Global Ozone Observing System, a programme of the World Meteorological Organization.

GOME Global Ozone Monitoring Experiment, a satellite system of the European Space Agency.

Greenhouse gas A GHG or greenhouse gas is a gaseous constituent of the atmosphere that absorbs and emits radiation within the spectrum of infrared radiation emitted by the Earth's surface, atmosphere and clouds.

GRIP The Greenland Ice-core Project of the European Union.

GURME Global Urban Research Meteorology and Environmental Project, a project of the World Meteorological Organization.

GWP Global Warming Potential, an index describing the radiative characteristics of a trace gas in terms of the time-weighted warming effect of unit mass emission of that trace gas compared with that of unit mass of emission of carbon dioxide.

Halon A compound whose structural formulae contain carbon, together with chlorine or bromine or fluorine or hydrogen and that may act as an ozone depleting agent.

HCFCs Hydrochlorofluorocarbons, compounds containing carbon, hydrogen, fluorine and chlorine that take part in stratospheric ozone depletion.

HMIP Her Majesty's Inspectorate of Pollution (United Kingdom).

IADN Integrated Atmospheric Deposition Network (United States of America).

ICP International Cooperative Program of the UNECE Convention on Long Range Transboundary Air Pollution.

IGAC International Global Atmospheric Chemistry project of the International Geosphere-Biosphere Programme.

IGBP International Geosphere-Biosphere Programme.

IGY International Geophysical Year, 1957.

IIASA International Institute for Applied Systems Analysis (Laxenburg, Austria).

IMECA Indice Metropolitano de la Calidad del Aire is an air quality index for Mexico.

IPCC Intergovernmental Panel on Climate Change, jointly established by the World Meteorological Organization and the United Nations Environment Programme.

ISKO A national air quality information system established in the Czech Republic.

Kyoto Protocol The Kyoto Protocol to the United Nations Framework Convention on Climate Change contains legally binding commitments to the reduction in emissions of greenhouse gases.

LAQN London Air Quality Network (United Kingdom).

LCPD Large Combustion Plant Directive, a regulation of the Commission of the European Communities.

Local scale This refers to the scale of a process or model that lies in the range from about 100 metres to about 10 kilometres.

LPG Liquified Petroleum Gases are mixtures of ethane, propane and butane that are by-products of oil refining, used for home heating and cooking.

LRTAP Long Range Transboundary Air Pollution.

MDN Mercury Deposition Network (United States of America).

Montreal Protocol The Montreal Protocol on Substances that Deplete the Ozone Layer was adopted in Montreal in 1987 and subsequently amended in London (1990), Copenhagen (1992), Vienna (1995), Montreal (1997) and Beijing (1999).

MOPITT Measurements of Pollution in the Troposphere, a satellite system carried by the NASA/Terra spacecraft.

MSA Methane sulphonic acid, $CH_3(SO_2)OH$, is an atmospheric degradation product of dimethyl sulphide.

MTBE Methyl t-butyl ether, $CH_3 O C(CH_3)_3$, is a petrol additive and anti-knock agent.

n-s SO_4 Non-sea salt sulphate is that component of particulate sulphate that does not arise from sea salts and sea spray.

NAAQOs National Ambient Air Quality Objectives (Canada).

NAAQS National Ambient Air Quality Standards (United States of America).

NADN National Atmospheric Deposition Network (United States of America).

NADP National Atmospheric Deposition Program (United States of America).

NAEI National Atmospheric Emissions Inventory (United Kingdom).

NAO North Atlantic Oscillation, is the fluctuation on timescales from days to decades in the pressure-difference between Iceland and the Azores.

NAPAP National Acid Precipitation Assessment Program is an activity on acidification in the United States of America.

NAPS National Air Pollution Surveillance network (Canada).

NASA National Aeronautics and Space Administration (United States of America).

NatChem National Atmospheric Chemistry data base (Canada).

NDSC Network for the Detection of Stratospheric Change.

NDVI Normalized Difference Vegetation Index, an index of the colour of the land surface of the Earth as seen by a satellite.

NECD National Emissions Ceilings Directive, a regulation of the Commission of the European Communities.

NETCEN National Environmental Technology Centre (Abingdon, United Kingdom).

NILU Norsk institutt for luftforskning (Kjeller, Norway).

Nimbus-7 A satellite launched by NASA in 1978.

NM-VOCs Non-methane volatile organic compounds, are organic compounds that act as tropospheric ozone precursors.

NOAA National Oceanic and Atmospheric Administration (United States of America).

NOx Oxides of nitrogen, denoted by the sum of the concentrations of nitric oxide (NO) and nitrogen dioxide (NO_2).

NTN National Trends Network (United States of America).

ODPs Ozone Depletion Potential is the extent of the depletion of the stratospheric ozone layer from the release of unit mass of that substance compared with that from unit mass of CFC-11, CCl_3F.

OECD Organization for Economic Cooperation and Development (Paris, France).

OH The hydroxyl OH radical is a highly reactive free radical species present in the atmosphere and responsible for the oxidation, degradation and removal of a wide range of trace gas species.

OTC Ozone Transport Commission. Working group established under the 1990 Clean Air Act Amendments (USA) to plan to meet air quality targets for ozone.

PAHs Polycyclic aromatic hydrocarbons, or polynuclear aromatic hydrocarbons, members of a large group of organic compounds whose molecular structures contain two or more aromatic rings fused together.

PCBs Polychlorinated biphenyls, members of a large group of organic compounds whose molecular structures contain two benzene rings directly joined and with hydrogen atoms replaced by chlorine atoms.

PEMEX Petroleos Mexicanos.

PFCs Perfluorocarbons, compounds whose molecular structures contain carbon and fluorine that may act as greenhouse gases.

pH This is an index used to express the hydrogen ion concentration of a solution and is defined as the common logarithm, with its sign reversed, of the hydrogen ion concentration expressed in gram ions per litre of solution.

PICCA Programa Integral contra la Contaminación Atmosférica (Integrated programme to combat atmospheric pollution).

PIRLA Paleoecological Investigation of Recent Lake Acidification, a joint programme between Canada and the United States of America.

PM Particulate matter, solid or liquid material suspended in the atmosphere and remaining airborne for at least several hours.

POPs Persistent Organic Pollutants.

Primary pollutants These are substances that are discharged directly into the atmosphere. Between the point of their discharge and the point at which they are removed, they cause damage to human health, crops, natural vegetation or to ecosystems.

PROAIRE Programa del Calidad del Aire (Programme to improve air quality).

PSCs Polar Stratospheric Clouds.

PSD Prevention of Significant Deterioration, a pollution control term used in the United States of America.

RAINS Regional Air Pollution Information System is an integrated assessment modelling tool designed for the development of policies to combat acidification, eutrophication and ground level ozone formation across Europe.

RAMA Red Automática de Monitereo Atmosférico (Automatic atmospheric monitoring network).

Regional scale This refers to the spatial scale of a process or a model that lies in the range from about 100 to about 10,000 kilometres.

RGAR Review Group on Acid Rain (United Kingdom).

RIVM Rijksinstituut voor Volksgezondheid en Mileu (Bilthoven, Netherlands).

S Sulphur.

SAR The Second Assessment Report of the Intergovernmental Panel on Climate Change.

SCAB South Coast Air Basin (California, United States of America).

SCAQMD South Coast Air Quality Management District (California, United States of America).

SCPs Spheroidal Carbonaceous Particles.

Secondary pollutants These are substances that are formed in the atmosphere by chemical reactions involving primary precursor pollutants. Between the point of their formation and the point of their ultimate removal, they cause damage to human health, crops, natural vegetation or to ecosystems.

SEDESOL An environmental body in Mexico.

SEDUE Secretaria de Desarollo Urbano y Ecologia, an environmental body in Mexico.

SEPA Scottish Environmental Protection Agency (United Kingdom).

SESMA Servicio de Salud del Ambiente Region Metropolitana is the environmental protection service for the Metropolitan Region of Mexico.

SIMAT Sistema de Monitoreo Atmosferico de la Ciudad de Mexico (Mexico).

SMB Simple Mass Balance model for acidification and critical loads assessment.

SPM Suspended Particulate Matter, solid or liquid material suspended in the atmosphere and remaining airborne for at least several hours.

SRES Special Report on Emission Scenarios, a publication of the Intergovernmental Panel on Climate Change.

SST Sea Surface Temperature.

SWAP Surface Water Acidification Programme, a joint programme between Sweden, Norway and the United Kingdom.

TAR The Third Assessment Report of the Intergovernmental Panel on Climate Change.

TEM Tropospheric Excess Method, is a methodology for determining the ozone content of the troposphere from two satellite measurements, one of which monitors the troposphere and stratosphere and the other which monitors the stratosphere alone.

TERRA This is the name of the NASA Earth Observing System satellite.

TOMPs Toxic Organic Micro Pollutants.

TOMS Total Ozone Mapping Spectrometer, a satellite-borne measurement system.

Total column ozone The total amount of ozone present in a column of air extending from the earth's surface to the top of the atmosphere. Most ozone usually occurs in the stratosphere.

TSP Total Suspended Particulates, solid or liquid material suspended in the atmosphere and remaining airborne for at least several hours.

UNECE United Nations Economic Commission for Europe.

UNEP United Nations Environment Programme.

UNFCCC United Nations Framework Convention on Climate Change has as its ultimate objective the stabilization of greenhouse gas concentrations in the atmosphere at a level that would prevent dangerous anthropogenic interference with the climate system.

US EPA United States Environmental Protection Agency (United States of America).

UV Ultraviolet radiation, is electromagnetic radiation with wavelengths beyond the violet end of the visible spectrum, that is below 400 nanometres.

UV-B Electromagnetic radiation with wavelengths in the range 280–320 nanometres.

VOCs Volatile organic compounds, are organic compounds that act as tropospheric ozone precursors.

WDC World Data Centre, a collaborating centre of the World Meteorological Organization.

WHO World Health Organization.

WMO World Meteorological Organization.

ZMCM Zona Metropolitana de la Ciudad de Mexico (Mexico).

Sources:
IPCC (2001).
WMO (2002).
Jerrard, H.G. and McNeill, D.B. 1992. *Dictionary of scientific units*. Chapman and Hall, London.

REFERENCES

Ahmed, A., Islam, K. and Reazuddin, M. 1996. An inventory of greenhouse gas emissions in Bangladesh: initial results. *Ambio* **25**, 300–303.

Alcamo, J., Mayerhofer, P., Guardans, R., van Harmelan, T., van Minnen, J., Onigkeit, J., Posch, M. and de Vries, B. 2002. An integrated assessment of regional air pollution and climate change in Europe: findings of the AIR-CLIM project. *Environmental Science and Policy* **5**, 257–272.

Ambio (2003) Biological recovery from acidification: Northern Lakes Recovery Study. *Ambio* **32** (3).

Ames, J., Myers, T.C., Reid, L.E., Whitney, D.C., Golding, S.H., Hayes, S.R. and Reynolds, S.D. 1985. *SAI airshed model operating manual. Vol. I: User's Manual.* US EPA Publication EPA-600/8-85-0007a, US EPA, Research Triangle Park, North Carolina, USA.

Andres, R.J., Marland, G., Fung, I. and Matthews, E. 1997. *Geographic patterns of carbon dioxide emissions from fossil-fuel burning, hydraulic cement production and gas flaring on a one degree by one degree grid cell basis: 1950 to 1990.* (NDP-058 at CDIAC).

Anfossi, D., Sandroni, S. and Viarengo, S. 1991. Tropospheric ozone in the nineteenth century: the Moncalieri series. *Journal of Geophysical Research* **96**, D9, 17,349–17,352.

Arndt, R., Carmichael, G., Streets, D. and Bhatti, N. 1997. Sulfur dioxide emissions and sectorial contributions to sulfur deposition in Asia. *Atmospheric Environment* **31**, 1553–1572.

Ashby, E. and Anderson, M. 1981. *The Politics of Clean Air.* Clarendon Press, Oxford.

Baggott, S.L., Davidson, I., Dore, C., Goodwin, J., Milne, R., Murrells, T.P., Rose, M., Watterson, J.D. and Underwood, B. 2003. *UK greenhouse gas inventory, 1990–2003.* AEAT/ENV/R/1396.

Barker, I. and Barker, J. 1988. *Clean Air around the World. The Law and Practice of air pollution control in 14 countries in 5 continents.* IAAPPA (International Association of Air Pollution Prevention Associations).

Barnola, J.M., Raynaud, D., Korotkevich, Y. and Lorius, C. 1987. Vostok ice core provides 160,000-year record of atmospheric CO_2. *Nature* **329**, 408–414.

Barrett, K., Schaug, J., Bartonova, A., Semb, A., Hjellbrekke, A-G. and Hanssen, J.E. 2000. *Europe's changing air environment. Two decades of trends in acidifying atmospheric sulphur and nitrogen in Europe 1978–1998.* EMEP/CCC-Report 7/2000.

Battarbee, R.W., Flower, R.J., Stevenson, A.C. and Rippey, B. 1985. Lake acidification in Galloway: a palaeoecological test of competing hypotheses. *Nature* **314**, 350–352.

Battarbee, R.W. and Charles, D.F. 1987. The use of diatom assemblages in lake sediments as a means of assessing the timing, trends and causes of lake acidification. *Progress in Physical Geography* **11**, 552–580.

Battarbee, R.W., Mason, J., Renberg, I. and Talling, J.F. eds. 1990. *Palaeolimnology and lake acidification.* The Royal Society.

Behrenfeld, M.J., Randerson, J.T., McClain, C.R., Feldman, G., Los, S.O., Tucker, C.J., Falkowski, P., Field, C., Frouin, R., Esaias, W., Kolber, D. and Pollack, N. 2001. Biospheric primary production during an ENSO transition. *Science* **291**, 2594–2597.

Benkovitz, C.M., Scholtz, M.T., Pacyna, J.M., Tarrason, L., Dignon, J., Voldner, E.C., Spiro, P.A., Logan, J.A. and Graedel. T.E. 1996. Global gridded inventories of anthropogenic emissions of sulfur and nitrogen. *Journal of Geophysical Research Atmospheres* **101**, D22, 29239–29253.

Berntsen, T. and Isaksen, I.S.A. 1997. A global three-dimensional chemical transport model for the troposphere. 1. Model description and CO and ozone results. *Journal of Geophysical Research* **102**, 21239–21280.

Bertrand, C., van Ypersele, J-P. and Berger, A. 1999. Volcanic and solar impacts on climate since 1700. *Climate Dynamics* **15**: 355–367.

Beverland, I. 1998. Urban air pollution and health. In: **J. Rose ed.** Environmental Toxicology – Current developments. *Environmental Topics*, vol. 7. Gorelon & Breach Science Publishers, pp. 189–209.

Bey, I., Jacob, D.J., Yantosca, R.M., Logan, J.A., Field, B.D., Fiore, A.M., Li, Q., Liu, H., Mickley, L.j. and Schultz, M.G. 2001. Global modeling of tropospheric chemistry with assimilated meteorology. Model description and evaluation. *Journal of Geophysical Research* **106**, 23073–23095.

Birks, H.J., Line, J.M., Juggins, S., Stevenson, A. and ter Braak, C.J. 1990. Diatoms and pH reconstruction. *Philosophical Transactions of the Royal Society of London* **B327**, 263–278.

Blake, D. and Sherwood Rowland, F. 1995. Urban leakage of liquified petroleum gas and its impact on Mexico City air quality. *Science* **269**, 953–956.

Boemare, C., Quirion, P. and Sorrell, S. 2003. The evolution of emissions trading in the EU: tensions between trading schemes and the proposed EU directive. *Climate Policy* 3S2, S105–S124.

Boer, G.J., Arpe, K., Blackburn, M., Deque, M., Gates, W.L., Hart, T.L., le Treut, H., Roeckner, H.E., Shenin, D.A., Simmonds, I., Smith, R.N.B., Tokioka, T., Wetherald, R.T. and Williamson, D. 1992. Some results from an intercomparison of climates simulated by 14 atmospheric general circulation models. *Journal of Geophysical Research* **97**, 12771–12786.

Borrell, P., Burrows, J.P., Platt, U. and Zehner, C. 2000. Determining tropospheric concentrations of trace gases from space. *European Space Agency Bulletin 2000*, TROPOSAT article. Available at http://www.crusoe.iup.uni-heidelberg.de/luftchem/troposat/articles/esabul-1.htm.

Borrell, P, Burrows, J.P., Richter, A., Platt, U. and Wagner, T. 2003. New directions: new developments in satellite capabilities for probing the chemistry of the troposphere. *Atmospheric Environment* **37**, 2567–2570.

Boutron, C.F., Candelone, J-P. and Hong, S. 1994. Past and recent changes in the large-scale tropospheric cycles of lead and other heavy metals as documented in Antarctic and Greenland snow and ice: a review. *Geochimica et Cosmochimica Acta* **58**, 3217–3225

Bouwman, A.F., Vanderhoek, K.W. and Olivier, J.G. 1995. Uncertainties in the global source distribution of nitrous oxide. *Journal of Geophysical Research*, **100**, D2, 2785–2800.

Bouwman, A.F., Lee, D.S., Asman, W.A., Dentener, F.J., VanderHoek, K.W. and Olivier, J.G. 1997. A global high-resolution emission inventory for ammonia. *Global Biogeochemical Cycles* **11**, 561–587.

Bouwman, A.F., Derwent, R.G. and Dentener, F.J. 1999. Towards reliable global bottom-up estimates of temporal and spatial patterns of emissions of trace gases and aerosols from land-use related and natural sources. In: A.F. Bouwman ed. Approaches to scaling factors of trace gases in ecosystems. *Elsevier Science*, pp. 3–26.

Bouwman, A.F., Van Vuuren, D.P., Derwent, R.G. and Posch, M. 2002. A global analysis of acidification and eutrophication of terrestrial ecosystems. *Water, Air and Soil Pollution* **141**, 349–382.

Bower, J. (1997) Ambient air quality monitoring. In: R.E. Hester and R.M. Harrison eds. Air Quality Management. *Issues in Environmental Science and Technology* **8**, 41–65. Royal Society of Chemistry.

Brasseur, G.P., Kiehl, J.T., Muller, J-F., Schneider, T., Granier, C., Tie, X.X. and Hauglustaine, D. 1998a. Past and future changes in global tropsopheric ozone: impact on radiative forcing. *Geophysical Research Letters* **25**, 3807–3810.

Brasseur, G.P., Hauglustaine, D., Walters, S., Rasch, P.J., Muller, J-F., Granier, C. and Tie, X.X. 1998b. MOZART, a global chemical transport model for ozone and related chemical tracers. *Journal Geophysical Research* **103**, 28265–28289.

Bravo, H.A. and Torres, R.J. 2000. The usefulness of air quality monitoring and air quality impact studies before the introduction of reformulated gasolines in developing countries. Mexico City, a real case study. *Atmospheric Environment* **34**, 499–506.

Brimblecombe, P. 1987. *The Big Smoke. A history of air pollution in London since Medieval times.* Methuen.

Brimblecombe, P. 1996. *Air composition and chemistry.* CUP.

Brimblecombe, P. ed. 2003. The effects of air pollution on the built environment. *Air Pollution Reviews*, vol. 2. Imperial College Press.

Brimblecombe, P. and Pitman, J. 1980. Long term deposit of Rothamsted, southern England. *Tellus* **32**, 261–267.

Broughton, G., Clark, H. and Willis, P. 2000. Air pollution in the UK: 1997. AEA Technology/NETCEN. AEAT 5303.

Burrows, J.P., Weber, M., Buchwitz, M., Rosanov, V., Ladstatter-Weissenmayer, A., Richter, A., DeBeek, R., Hogan, R., Bramstedt, K., Eichmann, K.U. and Eisinger, M. 1999. The global ozone monitoring experiment (GOME): mission concept and first scientific results. *Journal of Atmospheric Sciences* **56**, 151–175.

Bush, A. and Philander, S.G. 1999. The climate of the Last Glacial Maximum: results from a coupled atmosphere-ocean general circulation model. *Journal of Geophysical Research* **104**, D20, 24509–24525.

Butler, J.H., Battle, M., Bender, M.L., Montzka, S.A., Clarke, A.D., Saltzman, E.S., Sucher, C.M., Severinghaus, J. and Elkins, J. 1999. A record of atmospheric halocarbons during the twentieth century from polar firn air. *Nature* **399**, 749–755.

Cachier, H. 1995. Combustion carbonaceous aerosols in the atmosphere: implications for ice core studies. In: Ice core studies of global biogeochemical cycles. NATO ASI Ser. I, vol. 30, R.J. Delmas ed., pp.313–346. Springer Verlag.

Callendar, G. 1938. The artificial production of carbon dioxide and its influence on temperature. *Quarterly Journal of the Royal Meteorological Society* **64**, 223-....

CARB 1999. *The 1999 California almanac of emissions and air quality*. California Air Resources Board.

CARB 2003. *The 2003 California almanac of emissions and air quality*. California Air Resources Board. (http://www.arc.ca.gov/agd/almanac/almanac03/almanac03.htm)

Carmichael, G.R., Calori, G., Hayami, H., Uno, I., Cho, S.Y., Engardt, M., Kim, S-B., Ichikawa, Y., Ikeda, Y., Woo, J-H., Ueda, H. and Amann, M. 2002. The MICS-Asia study: model intercomparison of long-range transport and sulfur deposition in East Asia. *Atmospheric Environment* **36**, 175–199.

CEC 2000. The Auto-Oil II Programme: Air quality report. *Report of Working Group 1 on Environmental Objectives*, Directorate General for the Environment, Brussels. http://www.autooil.jrc.cec.eu.int/finalaq.htm.

Cerny, J. and Paces, T. eds. 1995. Acidification in the Black Triangle Region. Acid Reign 95 *5th International Conference on Acidic deposition, Science and Policy*, Goteborg, Sweden 1995. Excursion volume.

Chamberlain, A.C., Heard, M.J., Little, P., and Whiffen, R.D. 1979. The dispersion of lead from motor exhausts. *Philosophical Transactions of the Royal Society*, London. **A290**, 577–589.

Chang, J.S., Brost, R.A., Isaksen, I.S.A., Madronich, S., Middleton, P., Stockwell, W.R. and Walcek, C.J. 1987. A three-dimensional Eulerian acid deposition model: Physical concepts and formulation, *Journal of Geophysical Research* **92**, D12, 14681–14777.

Charlson, R.J. 1997. Direct climate forcing by anthropogenic sulfate aerosols: the Arrhenius paradigm a century later. *Ambio* **26**, 25–31.

Cifuentes, L., Borja-Aburto, V., Gouveia, N., Thurston, G. and Davis, D.L. 2001. Hidden health benefits of greenhouse gas mitigation. *Science* **293**, 1257–1259.

Cinderby, C., Cambridge, H., Herrera, R., Hicks, W., Kuylenstierna, J., Murray, F. and Olbrich, K. 1998. *Global assessment of ecosystem sensitivity to acidic deposition*. Stockholm Environment Institute, Stockholm.

City of Edinburgh Council 2001. Review and Assessment of Air Quality. Stage 3.

Civerolo, K.L., Brankov, E., Trivikrama Rao, S. and Zurbenko, I.G. 2001. Assessing the impact of the acid deposition control program. *Atmospheric Environment* **35**, 4135–4148.

CLAG 1994. *Critical loads of acidity in the United Kingdom*. DoE.

CLAG 1995. *Critical loads of acid deposition for United Kingdom Freshwaters*. CLAG Sub-group report on Freshwaters. DoE.

CLAG 1997. *Deposition fluxes of acidifying compounds in the United Kingdom*. CLAG Sub-group report on Deposition fluxes. DoE.

CLIMAP 1981. *Seasonal reconstructions of the earth's surface at the last glacial maximum*. Geological Society of America Map and Chart series, MC-36.

COHMAP members 1988. Climatic changes of the last 18,000 years: observations and model simulations. *Science* **241**, 1043–52.

Collins, M., Tett, S.F. and Cooper, C. 2001. The internal climate variability of the HadCM3, a version of the Hadley Centre coupled model without flux adjustments. *Climate Dynamics* **17**, 61–81.

Collins, W.J., Stevenson, D.S., Johnson, C.E. and Derwent, R.G. 1997. Tropospheric ozone in a global-scale three-dimensional Lagrangian model and its response to NO$_x$ emission controls. *Journal of Atmospheric Chemistry* **26**, 223–274.

Collins, W.J., Derwent, R.G., Johnson, C.E. and Stevenson, D.S. 2000. The impact of human activities on the photochemical production and destruction of tropospheric ozone. *Quarterly Journal of the Royal Meteorological Society* **126**, 1925–1951.

Colman, R.A. and McAvaney, B.J. 1995. Sensitivity of the climate response of an atmospheric general circulation model to changes in the convective parameterisation and horizontal resolution. *Journal of Geophysical Research* **100**, 3155–3172.

Colvile, R.N., Woodfield, N.K., Carruthers, D.J., Fisher, B.E., Rickard, A., Neville, S. and Hughes, A. 2002. Uncertainty in dispersion modeling and urban air quality mapping. *Environmental Science and Policy* **5**, 207–220.

COMEAP 1998. *The Quantification of the Effects of Air Pollution and Health in the United Kingdom*. HMSO.

Crawford, E. 1997. Arrhenius' 1896 model of the greenhouse effect in context. *Ambio* **26**, 6–11.

Crocker, T. 1966. The structuring of air pollution control systems. In: **H. Wolozin ed.** The economics of air pollution. **W.W. Norton**, pp. 61–86.

Crutzen, P.J., Lawrence, M.G. and Poschl, U. 1999. On the background photochemistry of tropospheric ozone. *Tellus* **51A-B**, 123–146.

Cubasch, U., Hasselmann, K., Hock, H., Maier-Reimer, E., Mikolajewicz, U., Santer, B.D. and Sausen, R. 1992. Time-dependent greenhouse warming – computations with a coupled ocean-atmosphere model. *Climate Dynamics* **8**, 55–69.

Czech Environmental Institute, eds 1999. *Statistical Environmental Yearbook of the Czech Republic 1999*.

DEFRA 2000a. *Review and Assessment: monitoring air quality*. LAQM.TG1(00).

DEFRA 2000b. *Review and Assessment: Estimating emissions*. LAQM. TG200 (both the above available from http://www.environment.detr.gov.uk/airq/laqm UPDATE)

DEFRA 2003 *The Air Quality Strategy for England, Scotland, Wales and Northern Ireland: Addendum.* DEFRA. PB7874.

De Leeuw, F., Moussiopoulos, N., Sahm, P. and Bartonova, A. 2001. Urban air quality in larger conurbations in the European Union. *Environmental Modelling and Software* **16**, 399–414.

Department of Health 1998. *Quantification of the effects of air pollution on health in the United Kingdom.* Committee on the Medical Effects of Air Pollutants. The Stationery Office.

Department of the Environment 1997. *The United Kingdom air quality strategy.* The Stationery Office.

Derwent, R.G., Middleton, D.R., Field, R.A., Goldstone, M.E., Lester, J.N. and Perry, R. 1995. Analysis and interpretation of air quality data from an urban roadside location in central London over the period from July 1991 to July 1992. *Atmospheric Environment* **29**, 923–946.

Derwent, R.G., Simmonds, P.G., O'Doherty, S. and Ryall, D.B. 1998. The impact of the Montreal protocol on halocarbon concentrations in northern hemisphere baseline and European air masses at Mace Head, Ireland over a ten year period from 1987 – 1996. *Atmospheric Environment* **32**, 3689–3702

Derwent, R.G., Stevenson, D.S., Collins, W.J. and Johnson, C.E. 2004. Intercontinental transport and the origins of ozone observed at surface sites in Europe. *Atmospheric Environment* **38**, 1891–1889.

DETR 1999. *Report on the Review of the National Air Quality Strategy Proposals to amend the Strategy.*

DETR 2000. *The Air Quality Strategy for England, Scotland, Wales and Northern Ireland.* Working together for clean air. Cm 4548. DETR.

Dlugokencky, E., Masarie, K., Lang, P. and Tans, P. 1998. Continuing decline in the growth rate of atmospheric methane burden. *Nature* **393**, 447–450.

Dockery, D.W., Pope, C.A., Xu, X., Spengler, J.D., Ware, J.H., Fay, M.E., Ferris, B.G. and Speizer, F.E. 1993. An association between air pollution and mortality in six US cities. *New England Journal of Medicine* **329**, 1753–1759.

DOE 1993. *Air pollution and tree health in the United Kingdom.* Report of the Terrestrial Effects Review Group. HMSO.

Dollard, G., Fowler, D., Smith, R.I., Hjellbrekke, A-G., Uhse, K. and Wallasch, M. 1995. Ozone measurements in Europe. *Water, Air and Soil Pollution* **85**, 1949–1954.

Dore, C.J., Goodwin, J.W., Watterson, J.D., Murrells, T.P., Passant, N.R., Hobson, M.M., Haigh, K.E., Baggott, S.L., Pye, S.T., Coleman, P.J. and King, K.R. 2003. *UK emissions of air pollutants 1970 to 2001.* (www.naei.co.uk/reports.php)

Doyle, R. 1997. Air pollution in the USA. *Scientific American* **276** (4), 27.

DUKES 2003. *Digest of United Kingdom Energy Statistics.* DTI.

EEA (European Environment Agency) 2001. *Air quality in larger cities in the European Union. A contribution to the Auto-Oil II Programme.* Topic report 3/2001. http://www.reports.eea.eu.int/Topic_report_No_032001/en.

EEA (European Environment Agency) 2004. *Analysis of greenhouse gas emission trends and projections in Europe 2003.* EEA Technical Report 4/2004.

Eggleston, S., Hackman, M., Heyes, C., Irwin, J., Timmis, R. and Williams, M. 1992. Trends in urban air pollution in the United Kingdom during recent decades. *Atmospheric Environment* **26B**, 227–239.

EIA 2003. *International energy outlook 2003.* Energy Information Administration. DOE/EIA-0484(2003).

Eliassen, A. and Saltbones, J. 1983. Modelling the long-range transport of sulphur over Europe: a two year model run and some model experiments. *Atmospheric Environment* **17**, 1457–1473.

Elsom, D. 1996. *Smog Alert. Managing urban air quality.* Earthscan.

EMEP/CORINAIR 2004. *Emissions inventory guidebook.* 3rd Edition, September 2003 update. European Environment Agency Technical Report No. 30.

Emori, S., Nozawa, T., Abe-Ouchi, A., Numaguti, A. and Kimoto, M. 1999. Coupled ocean-atmosphere model experiments of future climate change with an explicit representation of sulphate aerosol scattering. *Journal of the Meteorological Society* Japan **77**, 1299–1307.

Engstrom, D. and Swain, E. 1997. Recent declines in atmospheric mercury in the Upper Midwest. *Environmental Science and Technology* **31**, 960–967.

Environment Canada 1997. *1997 Canadian Acid Rain Assessment.*

Environment Canada 2001. *Environmental Protection Series, National Air Pollution Surveillance Network.* Annual Summary for 1999. EPS 7/AP/32

EPA 1992. *National air pollutant emission estimates 1900–1991.* EPA Publication No. EPA-454/R-92-013.

EPA 1998. *National air quality and emissions trends report, 1997.* EPA 454/R-98-016.

EPA 1999. *Inventory of US Greenhouse gas emissions and sinks: 1990–1997.* US EPA 236-R-99-003.

EPA 2001. *The United States experience with economic incentives for protecting the environment.* EPA-240-R-01-001.

EPA 2002. *Clearing the air. The facts about capping and trading emissions.* EPA-430F-02-009.

EPA 2003a. *US emission inventory 2003. Inventory of US greenhouse gas emissions and sinks: 1990–2001.* EPA 430-R-03-004.

EPA 2003b. *Latest findings on national air quality. 2002 status and trends.* US EPA 454/k-03-001.

EPICA Community members 2004. Eight glacial cycles from an Antarctic ice core. *Nature* **429**, 623–628.

European Environment Agency (EEA) 2000. *Environmental signals 2000.* Environmental Assessment report no. 6.

European Environment Agency (EEA) 2003. *Greenhouse gas emission trends and projections in Europe 2003. Tracking progress by the EU and acceding and candidate countries towards achieving their Kyoto Protocol targets.* Environmental Issues Report 36/2003.

Farman, J.C., Gardiner, B.G. and Shanklin, J.D. 1985. Large losses of total ozone in Antarctica reveal seasonal ClO_x/NO_x interaction. *Nature* **315**, 207–210.

Farmer, J.G. 1994. Environmental change and the chemical record in Loch Lomond sediments. *Hydrobiologia* **290**, 39–49.

Fenger, J. 1999. Urban air quality. *Atmospheric Environment* **33**, 4877–4900.

Ferrari, C., Clotteau, T., Thompson, L.G., Barbante, C., Cozzi, G., Cescon, P., Hong, S., Maurice-Bourgin, L., Francou, B. and Boutron, C. 2001. Heavy metals in ancient tropical ice: initial results. *Atmospheric Environment* **35**, 5809–5815.

Fiala, J. Cernikovsky, L., de Leeuw, F. and Kurfuerst, P. 2003. *Air pollution by ozone in Europe in summer 2003.* European Environment Agency Topic Report 3/2003.

Finlayson-Pitts, B.J. and Pitts, Jr., J.N. 1999. *Chemistry of the upper and lower atmosphere.* Academic Press.

Fishman, J., Ramanathan, V., Crutzen, P.J. and Liu, S.C. 1979. Tropospheric ozone and climate. *Nature* **282**, 818–820.

Fishman, J, Brackett, V.G., Browell, E.V. and Grant, W.B. 1996. Tropospheric ozone derived from TOMS/SBUV measurements during TRACEA. *Journal of Geophysical Research – Atmospheres*, **101**, D19, 24,069-24,082.

Fitzgerald, W.F, Engstrom, D.R, Mason, R.P. and Nater, E.A. 1997. The case for atmospheric mercury contamination in remote areas. *Environmental Science and Technology* **32**, 1–7.

Gagnon, C. 2001. *Rapport annuel de la qualité de l'air 2000.* Communauté Urbaine de Montréal. Service de l'environement.

Galpin, P.F., van Doormal, J.P. and Railby, G.D. 1985. Solution of the incompressible mass and momentum equations by application of a coupled line solver. *International Journal of Numerical Methods for Fluids* **5**, 615–625.

Garnett, A. 1967. Some climatological problems in urban geography with reference to air pollution. *Transactions of the Institute of British Geographers* **42**, 21–43.

GDF (Gobierno del Distrito Federal) 2003. *Informe del estado de la calidad del aire y tendencias 2002.* ZMVM. (http://www.sma.df.gob.mx/simat/pninfoanual.htm)

GEMS/AIR 1996. *Air Quality Management and Assessment Capabilities in 20 major cities.* UNEP/DEIA/AR.96.2; WHO/EOS 95.7. MARC, London.

Gery, M.W. and Crouse, R. 1990. *User's guide for executing OZIPR. Final Report.* United States Environmental Protection Agency, Research Triangle Park, North Carolina, USA.

Gillies, D. 1999. *A Guide to EC Environmental Law.* Earthscan.

Goettsch, W., Garssen, J., Slob, W., de Gruijl, F. and Van Loveren, H. 1998. Risk assessment for the harmful effects of UVB radiation on the immunological resistance to infectious diseases. *Environmental Health Perspectives* **106**, 71–77.

Goodwin, J., Salway, A.G., Eggleston, H.S, Murrells, T.P. and Berry, J.E. 1999. *UK emissions of air pollutants 1970 to 1996.* National Atmospheric Emissions Inventory, AEA Technology.

Goodwin, J.W., Salway, A.G., Dore, C.J., Murrells, T.P., Passant, N.R., Watterson, J.D., Hobson, M.M., Haigh, K.E., King, K.R., Pye, S.T., Coleman, P.J. and Conolly, C.M. 2002. *UK emissions of air pollutants 1970–2000.* NETCEN/AEA Technology Report.

Government of Canada 2002. *Climate change plan for Canada.* (Available at http://www.climatechange.gc.ca)

Grinder, R.D. 1980. *The battle for clean air: the smoke problem in post-Civil War America.* In: *Pollution and reform in American cities, 1870–1930.* **Melosi, M.V.** ed., University of Texas Press, pp. 83–103.

Gupta, J., Olsthoorn, X. and Rotenberg, E. 2003. The role of scientific uncertainty in compliance with the Kyoto Protocol to the Climate Change Convention. *Environmental Science and Policy* **6**, 475–486.

Haagen-Smit, A.J., Bradley, C.M. and Fox, M.M. 1953. Ozone formation in photochemical oxidation of organic substances. *Industrial Engineering Chemistry* **45**, 2080–2089.

Hammer, C.V. 1977. Past volcanism revealed by Greenland ice sheet impurities. *Nature* **270**, 482–486.

Hammer, C, Clausen, H. and Dansgaard, W. 1980. Greenland ice sheet evidence of post-glacial volcanism and its climatic impact. *Nature* **288**, 230–235.

Hannegan, B., Olsen, S., Prather, M., Zhu, X., Rind, D. and Lerner, J. 1998. The dry stratosphere: A limit on cometary water influx. *Geophysical Research Letters* **25**, 1649–1652.

Hansen, J.M., Sato, M. and Ruedy, R. 1997. Radiative forcing and climate response. *Journal of Geophysical Research* **102**, 6831–6864.

Harrison, S.P., Kohfeld, K., Roelandt, C. and Claquin, T. 2001. The role of dust in climate changes today, at the last glacial maximum and in the future. *Earth Science Reviews* **54**, 43–80.

Henriksen, A., Skjevale, B.L., Mannio, J., Wilander, A., Harriman, R., Curtis, C., Jensen, J.P., Fjeld, E. and Moiseenko, T. 1998. Northern European Lake Survey 1995. *Ambio* **27**, 80–91.

Heyes, C., Schopp, W., Amann, M. and Unger, S. 1996. *A reduced-form model to predict long-term ozone concentrations in Europe.* IIASA Interim Report WP-96-12/December.

Hill, M.R. 1997. Energy, environment and technology in the former USSR: the case of fossil fuelled power stations and acid rain emissions. *Energy and Environment* **8**, 247–267.

Hill, M.R. 1999. Carbon dioxide emissions from the Russian Federation – problems and choices. *Energy and Environment* **10**, 51–78.

Holdsworth, G. and Peake, E. 1985. Acid content of snow from a mid-troposphere sampling site on Mount Logan, Yukon territory, Canada. *Annals of Glaciology* **7**, 153–160.

Hong, S., Candelone, J-P., Patterson, C. and Boutron, C. 1994. Greenland ice evidence of hemispheric lead pollution by Greek and Roman civilisations. *Science* **265**, 1841–1843.

Hong, S., Candelone, J-P. and Boutron, C.F. 1997. Changes in zinc and cadmium concentrations in Greenland ice during the past 7760 years. *Atmospheric Environment*, **31**, 2235–2242.

Hongmin, D., Erda, L., Yue, L., Minjie, R. and Qichang, Y. 1996. An estimation of methane emissions from agricultural activities in China. *Ambio* **25**, 292–296.

Hough, A.M. and Derwent, R.G. 1990. Changes in the global concentration of tropospheric ozone due to human activities. *Nature* **344**, 645–648.

Houghton, J.T., Jenkins, G.J and Ephraums, J.J. (eds) 1990. *Climate change: the IPCC scientific assessment.* CUP.

Houghton, J. T., Gylvan Meira Filho, L., Griggs, D.J. and Maskell, K. 1997. *An introduction to simple climate models used in the IPCC second assessment report.*

Hrbáček, J., Binek, B. and Mejstřík, V. 1989. Czechoslovakia. **Kurmondy, E.J. ed.** *International Handbook of Pollution Control*, pp. 137–151. Greenwood Press.

Instituto Nacional de Ecologia 2003. http://www/ ine.gob.mx/dgicurg/calaire/lineas/tendencias/zmvm/ ozono/index.html.

INEGI 1998. *Estadisticas del Medio Ambiente, Mexico 1997.* Instituto Nacional de Estadística, Geografia e Informática, Aquascalientes.

IPCC *Technical Paper II.* WMO/UNEP.

IPCC 1990. *Climate change. The IPCC Scientific assessment.* **J.T. Houghton et al. eds.** Cambridge University Press.

IPCC 1995. *Climate change 1994. Radiative forcing of climate change and an evaluation of the IPCC IS92 emission scenarios.* Cambridge University Press.

IPCC Working Group I 1996. *Climate change 1995. The Science of climate change.* **J.T. Houghton et al. eds**. Cambridge University Press.

IPCC 1996. *Revised 1996 IPCC Guidelines for National greenhouse gas inventories.* **J.T. Houghton et al. eds.** IPCC/OECD/IEA.

IPCC Working Group III 2000. *Special report on emission scenarios. Summary for Policy Makers.*

IPCC Working Group I 2001. *Climate change 2001. The Scientific basis.* **J.T. Houghton et al., eds.** Cambridge University Press.

Jenkins, A., Ferrier, R.C. and Wright, R.F. 2001. Assessment of recovery of European surface waters from acidification 1970– 2000. *Hydrology and Earth System Sciences* **5** (3). Special issue.

Jeuken, A.B., Eskes, H.J., van Velthoven, P.F., Kelder, H.M. and Holm, E.U. 1999. Assimilation of total ozone satellite measurements in a three-dimensional tracer transport model. *Journal of Geophysical Research* **104**, 5551–5563.

Jones. A.P. 1999. Indoor air quality and health. *Atmospheric Environment* **33**, 4535–4564.

Jones, P.D. Parker, D.E., Osborn, T.J. and Briffa, K. 2001. *Global and hemispheric temperature anomalies – land and marine instrumental.* In *Trends: a compendium of data on global change.* CDIAC, Oak Ridge National Laboratory, US Department of Energy, Oak Ridge, USA. (http:// www.cdiac.esd. ornl.gov/trends/trends.htm)

Jorquera, H. 2002a. Air quality in Santiago, Chile: a box modelling approach – I. Carbon monoxide, nitrogen oxides and sulfur dioxide. *Atmospheric Environment* **36**, 315–330.

Jorquera, H. 2002b. Air quality in Santiago, Chile: a box modelling approach II. $PM_{2.5}$, coarse and PM_{10} particulate matter fractions. *Atmospheric Environment* **36**, 331–344.

Kato, N. and Akimoto, H. 1992. Anthropogenic emissions of SO_2 and NO_x in Asia: emission inventories. *Atmospheric Environment* **26A**, 2997–3017.

Keeling, C.D., Whorf, T.P., Wahlen, M. and van der Plicht, J. 1995. Interannual extremes in the rate of rise of atmospheric carbon dioxide since 1980. *Nature* **375**, 666–670.

Keeling, C.D. and Whorf, T.P. 2003. *Atmospheric CO_2 records from sites in the SIO air sampling network.* In *Trends: a compendium of data on global change.* Carbon Dioxide Information Analysis Center, Oak Ridge National Laboratory, USA.

Klug, W., Grazianai, G., Grippa, G., and Tassone, C. 1990. *The ATMES report. Evaluation of long range atmospheric transport models using environmental radioactivity data from the Chernobyl accident.* Elsevier Scientific Publishing Company, London.

Knutson, T.R., Delworth, T.L., Dixon, K.W., and Stouffer, R.J. 1999. Model assessment of regional surface temperature trends (1947–97). *Journal of Geophysical Research* **104**, 30981–30996.

Kohfeld, K.E. and Harrison, S.P. 2000. How well can we simulate past climates? Evaluating the models using global palaeoenvironmental datasets. *Quaternary Science Reviews* **19**, 321–346.

Krupnick, A. and Sebastian, I. 1990. *Issues in urban air pollution: a review of the Beijing case study.* World Bank.

Kutzbach, J.E. 1985. Modeling of paleoclimates. *Advances in Geophysics* **28A**, 159–196.

Lacy, R. 1993. *La calidad del aire en al Valle de Mexico.* El Colegio de Mexico.

Lamb, D. and Bowersox, V. 2000. The National Atmospheric Deposition Program: an overview. *Atmospheric Environment* **34**, 1661–1663.

Lamb, H.H. 1970. Volcanic dust in the atmosphere: with a chronology and assessment of its meteorological significance. *Philosophical Transactions of the Royal Society of London* **A266**, 425–533.

Law, K.S., Plantevin, P-H., Shallcross, D.E., Rogers, H.L., Pyle, J.A., Groubel, C., Thouret, V. and Marenco, A. 1998. Evaluation of modelled O_3 using MOZAIC data. *Journal of Geophysical Research* **103**, 25721–25740.

Laxen, D.P. and Thompson, M. 1987. Sulphur dioxide in Greater London 1931–1985. *Environmental Pollution* **43**, 103–114.

Lee, D.S., Kohler, I., Grobler, E., Rohrer, F., Sauen, R., Gallardo-Klenner, L., Olivier, J., Dentener, F. and Bouman, A.F. 1997. Estimates of global NO_x emissions and their uncertainties. *Atmospheric Environment* **31**, 1735–1749.

Legrand, M., Feniet-Saigne, C., Saltzman, E., Germain, C., Barkov, N. and Petrov, V. 1991. Ice-core record of oceanic emissions of dimethylsulphide during the last climate cycle. *Nature* **350**, 144–146.

Lindqvist, O., Johansson, K., Aastrup, M., Andersson, A., Bringmark, L., Housenius, G., Hakanson, L., Iverfeldt, A., Meili, M. and Timm, B. 1991. Mercury in the Swedish environment. Recent research on causes, consequences and corrective methods. *Water, Air and Soil Pollution* **55** (1–2). Special issue.

Lloyd, A.C. 1997. *California's approach to air quality management*. In: *Issues in Environmental Science and Technology 8*. **R.E. Hester and R.M. Harrison eds.** 141–156.

Lomborg, B. 2001. *The skeptical environmentalist: measuring the real state of the world*. Cambridge University Press.

Lorius, C., Jouzel, J. Raynaud, D., Hansen, J. and Le Treut, H. 1990. *The ice-core record: climate sensitivity and future greenhouse warming*.

Lovelock, J.E. 1972. Atmospheric turbidity and CCl3F concentrations in rural southern England and southern Ireland. *Atmospheric Environment* **6**, 917–925.

Manabe, S. and Broccoli, A.J. 1985. The influence of continental ice sheets on the climate of the ice age. *Journal of Geophysical Research* **90**, C2, 2167–2190.

Manabe, S., Stouffer, R.J., Spelman, M.J. and Bryan, K. 1991. Transient responses of a coupled ocean-atmosphere model to gradual changes of atmospheric CO_2. Part I: annual mean response. *Journal of Climate* **4**, 785–818.

Marland, G., Boden, T.A. and Andres, R.J. 2003. *Global, regional and national CO_2 emissions*. In *Trends: a compendium of data on global change*. CDIAC.

Mason, R.P., Fitzgerald, W.F. and Morel, F.M. 1994. The biogeochemical cycling of elemental mercury: anthropogenic influences. *Geochimica et Cosmochimica Acta* **58**, 3191–3198

Mayewski, P.A. and Legrand, M.R. 1990. Recent increase in nitrate concentration of Antarctic snow. *Nature* **346**, 258–260.

Mayewski, P.A., Lyons, W.B., Spencer, M.J., Twickler, M.S., Buck, C.F. and Whitlow, S. 1990. An ice core record of atmospheric response to anthropogenic sulphate and nitrate. *Nature* **346**, 554–556.

McRae, G.J., Goodin, W.R. and Seinfeld, J.H. 1982. Numerical solution of the atmospheric diffusion equation for chemically reacting flows. *Journal of Computational Physics* **45**, 1–42.

Meehl, G.J., Boer, G.J., Covey, C., Latif, M., and Stouffer, R.J. 2000. The Coupled Model Intercomparison Project (CMIP). *Bulletin of the American Meteorological Society* **81**, 313–318.

Melosi, M. 1980. *Environmental crisis in the city: the relationship between industrialisation and urban pollution*. In: *Pollution and Reform in American Cities 1870–1930*. **M.L. Melosi (ed).** University of Texas Press.

Mickley, L.J., Murti, P.P., Jacob, D.J., Logan, J.A., Rind, D. and Koch, D. 1999. Radiative forcing from tropospheric ozone calculated with a unified chemistry-climate model. *Journal of Geophysical Research* **104**, 30153–30172.

Missfeldt, F. and Villavicenco, A. 2000. The economies in transition as part of the climate regime: recent developments. *Environment and Planning* B **27**, 379–392.

Mitchell, J.F.B., Grahame, N.S. and Needham, K.J. 1988. Climate simulations for 9000 BP: seasonal variations and effect of the Laurentide ice sheet. *Journal of Geophysical Research* **93**, D7, 8283–8303.

Mitchell, J.F.B., Johns, T.C., Gregory, J.M. and Tett, S.F. 1995. Climate response to increasing levels of greenhouse gases and sulphate aerosols. *Nature* **376**, 501–504.

Mitchell, J.M. 1970. *A preliminary evaluation of atmospheric pollution as a cause of global temperature fluctuation of the past century*. In: *Global Effects of Environmental Pollution*, **S.F. Singer and D. Reidel eds.**, 139–155.

Molina, L.T. and Molina, M.J. 2002. *Air quality in the Mexico megacity: an integrated assessment*. Kluwer Academic.

Molina, M.J. and Rowland, F.S. 1974. Stratospheric sink for chlorofluoromethanes: chlorine atom-catalysed destruction of ozone. *Nature* **249**, 810–812.

Monteith, D.T. and Evans, C.D. eds. 2000. *UK Acid waters monitoring network: 10 year report*. ENSIS Publishing.

Moussiopoulos, N., Berge, E., Bøhler, T., de Leeuw, F., Grønskei, K-E., Mylona, S. and Tombrou, M. 1996. *Ambient air quality, pollutant dispersion and transport models*. Topic report 19/96. European Topic Centre on Air Quality. http://www.reports.eea.eu.int/92-9167-028-6/en/tab_content_RLR.

Moussiopoulos, N., Borrego, C., Bozo, L., Galmarini, S., Poppe, D., Schatzmann, M. and Sturm, P. 2003. *Urban and local scale air pollution*. In: *Towards Cleaner Air for Europe – Science, Tools and Applications*. **Eds. P. Midgley and M. Reuther**. Margraf Publishers, Leiden, The Netherlands.

Mulholland, M. and Seinfeld, J. 1995. Inverse air pollution modelling of urban scale carbon monoxide emissions. *Atmospheric Environment* **29**, 497–516.

Muller, J-F. and Brasseur, G. 1995. IMAGES: A three-dimensional chemical transport model of the global troposphere. *Journal of Geophysical Research* **100**, 16445–16490.

Murley, L.ed. 1995. *Clean Air around the World.* IUAPPA. 3rd edition.

Mylona, S. 1996. Sulphur dioxide emissions in Europe 1880 – 1991 and their effects on sulphur concentrations and depositions. *Tellus* **48B**, 662–689.

Nath, B. 1997. *EU environmental legislation and the 'six': past, present and future.* Proceedings of the International Conference on EU Directives and Environmental Law, Prague 1997, 33–45.

National Research Council 1991. *Rethinking the ozone problem.* National Academy Press, Washington DC, USA.

Neftel, A., Beer, J., Oeschger, H., Zurcher, F. and Finkel, R.C. 1985. Sulphate and nitrate concentrations in snow from South Greenland 1895–1978. *Nature* **314**, 611–613.

NEGTAP 2001. *Transboundary air pollution: acidification, eutrophication and ground-level ozone in the UK.* National Expert Group on Transboundary Air Pollution.

Ng, A. and Patterson, C.C. 1981. Natural concentrations of lead in ancient Arctic and Antarctic ice. *Geochimica et Cosmochimica Acta* **45**, 2109–2121.

Nilsson, J. and Grennfelt, P. eds. 1988. *Critical loads for sulphur and nitrogen.* Report of the Skokloster workshop. Miljørapport 15. Nordic Council of Ministers, Copenhagen.

Noda, A., Yoshimatsu, K., Kitoh, A. and Koide, H. 1999. *Relationship between natural variability and CO_2-induced warming pattern: MRI coupled atmosphere /mixed-layer (slab) ocean GCM (SGCM) experiment.* 10th Symposium on Global Change Studies, pp 353–358, American Meteorological Society, Boston, Mass. USA.

Norwegian State Pollution Control Authority 1987. *1000 Lake Survey 1986 Norway.* Report 283/87.

Nozawa, T., Emori, S., Numagati, A., Tsuhima, Y., Takemura, T., Nakajima, T., Abe-Ouchi, A. and Kimoto, M. 2001. *Projections of future climate change in the 21st century simulated by the CCSR/NIES CGCM under the IPCC SRES scenarios.* In: *Present and Future of Modelling Global Environmental Change – Toward Integrated Modelling.* **Ed. T. Matsuno.** Terra Scientific Publishing Company, Tokyo, Japan.

Nriagu, J.O. 1991. *Human influence on the global cycling of trace metals.* In: *Heavy Metals in the Environment,* **J.G. Farmer (ed)**, International Conference Edinburgh 1991, 1–5.

OECD 1996. *Environmental information systems in the Russian Federation, an OECD assessment.* Paris.

OECD 1999. *Advanced air quality indicators and reporting. Methodological study and assessment.* ENV/EPOC/PPC(99)9/FINAL.

Palutikof, J. 2003. 1. *Global temperature record. CRU Information Sheet no. 1.* Climatic Research Unit, University of East Anglia. (http://www.cru.ac.uk/cru/info/warming/)

Passant, N. 2003. *Estimation of uncertainty in the National Atmospheric Emissions Inventory.* AEAT/ENV/R/0139 issue 1.

Patrick, S., Monteith, D.T. and Jenkins, A. 1995. *UK Acid waters monitoring network: the first five years. Analysis and interpretation of results, April 1988 – March 1993.* ENSIS publishing for the Department of the Environment.

Petersen, G., Iverfeldt, A. and Munthe, J. 1995. Atmospheric mercury species over central and northern Europe. Model calculations and comparison with observations from the Nordic air and precipitation composition network for 1987 and 1988. *Atmospheric Environment* **29**, 47–67.

Peterson, D.J. 1993. *Troubled Lands. The legacy of Soviet environmental destruction.* Westview.

Petit, J.R., Mounier, L., Jouzel, J. Korotkevich, Y.S., Kotlyakov, V.I. and Lorius, C. 1990. Palaeoclimatological and chronological implications of the Vostok core dust record. *Nature* **343**, 56–58.

Petit, J.R., Jouzel, J., Raynaud, D., Barkov, N., Barnola, J-M., Basile, I., Benders, M., Chappellaz, J., Davis, M., Delaygue, G., Delmotte, M., Kotlyakov, V., Legrand, M., Lipenkov, V, Lorius, C., Pepin, L., Ritz, C., Saltzman, E. and Stievenard, M. 1999. Climate and atmospheric history of the past 420,000 years from the Vostok ice core, Antarctica. *Nature* **399**, 429–436.

Petit, J.R. et al. 2001. *Vostok ice core data for 420,000 years.* IGBP PAGES/WDC for Paleoclimatology, DCS #2001-076. NOAA/NGDC Paleoclimatology Program, Boulder. http://www.ngdc.noaa.gov/paleo/forcing.html#tracegas.

Pitari, G., Grassi, B. and Visconti, G. 1997. *Results of a chemical-transport model with interactive aerosol microphysics.* Proc. XVIII Quadrennial Ozone Symposium. **Eds: R. Bojkov and G. Visconti,** pp. 759–762.

PORG 1997. *Ozone in the United Kingdom.* 4th Report of the Photochemical Oxidants Review Group.

Prescott, G.J., Cohen, G.R., Elton, R.A., Fowkes, F.G. and Agius, R. 1998. Urban air pollution and cardiopulmonary ill health: a 14.5 year time series study. *Occupational and Environmental Medicine* **55**, 697–704.

Prinn, R.G., Huang, J., Weiss, R.F., Cunnold, D.M., Fraser, P.J., Simmonds, P.G., McCulloch, A., Harth, C., Salameh, P., O'Doherty, S., Wang, R.H., Porter, L. and Miller, B. 2001. Evidence for substantial variations of hydroxyl radicals in the past two decades. *Science* **292**, 1882–1888.

QUARG 1993. *Urban air quality in the United Kingdom.* First report of the Quality of Urban Air Review Group. DETR.

Ramstein, G., Serafini-le Treut, Y., le Treut, H., Forichon, M. and Joussaume, S. 1998. Cloud processes associated with past and future climate changes. *Climate Dynamics* **14**, 233–247.

Renberg, I. 1986. Concentration and annual accumulation values of heavy metals in lake sediments: their significance in studies of the history of heavy metal pollution. *Hydrobiologia* **143**, 379–385.

Renberg, I., Persson, M.W. and Emteryd, O. 1994. Preindustrial atmospheric lead contamination detected in Swedish lake sediments. *Nature* **368**, 323–326.

RGAR 1983. *Acid Deposition in the United Kingdom*. The UK Review Group on Acid Rain. Warren Spring Laboratory.

RGAR 1990. *Acid Deposition in the United Kingdom 1986 – 1988*. 3rd Report of the UK Review Group on Acid Rain.

RGAR 1997. *Acid Deposition in the United Kingdom 1992 – 1994*. 4th Report of the UK Review Group on Acid Rain.

Robok, A. and Mao, J. 1995. The volcanic signal in surface temperature observations. *Journal of Climate* **8**, 1086–1103.

Rodhe, H. and Granat, L. 1984. An evaluation of sulfate in European precipitation 1955–1982. *Atmospheric Environment* **18**, 2627–2639.

Rodhe, H. and Charleson, R. eds. 1997. Svante Arrhenius and the Greenhouse effect. *Ambio* 26 (1). Special issue.

Romero, H., Ihl, M., Rivera, A., Zalazar, P. and Azocar, P. 1999. Rapid urban growth, land-use changes and air pollution in Santiago, Chile. *Atmospheric Environment* **33**, 4039–4047.

Rose, N., Juggins, S. and Watt, J. 1996. Fuel-type characterization of carbonaceous fly-ash particles using EDS-derived surface chemistries and its application to particles extracted from lake sediments. *Proceedings of the Royal Society of London* **452**, 881–907.

Rosman, K., Chisholm, W., Hong, S., Candelone, J-P. and Boutron, C.F. 1997. Lead from Carthaginian and Roman Spanish mines isotopically identified in Greenland ice dated from 600 BC to 300 AD. *Environmental Science and Technology* **31**, 3413–3416.

Russian Federation 2002. *3rd National communication of the Russian Federation*. Submitted in accordance with articles 4 and 12 of the UNFCCC. Inter-agency Commission of the Russian Federation on Climate Change.

Ryaboshapko, A., Ilyin, I., Gusev, A., Afinogenova, O., Berg, T. and Hjellbrekke, A-G. 1999. *Monitoring and modelling of lead, cadmium and mercury transboundary transport in the atmosphere of Europe*. EMEP/MSC-E Report 3/99.

Ryerson, T.B., Trainer, M., Angevine, W.M., Brock, C.A., Dissly, R.W., Fehsenfeld, F.C., Frost, G.J., Goldan, P.D., Holloway, J.S., Hubler, G., Jakoubek, R.O., Kuster, W.C., Neuman, J.A., Nicks, D.K., Parrish, D.D., Roberts, J.M., Sueper, D.T., Atlas, E.L., Donnelly, S.G., Flocke, F., Fried, A., Potter, W.T., Schauffler, S., Stroud, V., Weinheimer, A.J., Wert, B.P., Wiedinmyer, C., Alvarez, R.J., Banta, R.M., Darby, L.S. and Senff, C.J. 2003. Effect of petrochemical industrial emissions of reactive alkenes and NOx on tropospheric ozone formation in Houston, Texas. *Journal of Geophysical Research* **108**, 4249, 10.1029/2002JD003070.

Schopp, W., Amann, M., Cofala, J., Heyes, C. and Klimont, Z. 1999. Integrated assessment of European air pollution emission control strategies. *Environmental Modelling and Software* **14**, 1–9.

Seigneur, C., Karamchandani, P., Lohman, K. and Vijayaraghavan, K. 2001. Multiscale modeling of the atmospheric fate and transport of mercury. *Journal of Geophysical Research* **106**, D21, 27,795–27,809.

Seinfeld, J.H. and Pandis, S.N. 1998. *Atmospheric chemistry and physics. From air pollution to climate change*. Wiley Interscience.

Senior, C.A. and Mitchell, J.F.B. 2000. The time-dependence of climate sensitivity. *Geophysical Research Letters* **27**, 2685–2688.

Severinghaus, J.P. and Brook, E.J. 1999. Abrupt climate change at the end of the last glacial period inferred from trapped air in polar ice. *Science* **286**, 930–934.

Shahgedanova, M. and Burt, T.P. 1994. New data on air pollution in the former Soviet Union. *Global Environmental Change* 4, 201–227.

Shahgedanova, M., Burt, T.P. and Davies, T.D. 1999. Carbon monoxide and nitrogen oxides pollution in Moscow. *Water, Air and Soil Pollution* **112**, 107–131.

Shaw, D.J. and Oldfield, J. 1998. The natural environment of the CIS in the transition from communism. *Post-Soviet Geography and Economics* **39**, 164–177.

Shi Zongbo, Shao Longyi, Jones, T.P., Whiottaker, A.G., Lu Senlin, Berube, K.A., He Taoe and Richards, R.J. 2003. Characterization of airborne individual particles collected in an urban area, a satellite city and a clean air area in Beijing, 2001. *Atmospheric Environment* **37**, 4097–4108.

Shine, K.P. and Forster, P.M. de F. 1999. The effects of human activity on radiative forcing of climate change: a review of recent developments. *Global and Planetary Change* **20**, 205–225.

Simpson, D. 1992. Long period modelling of photochemical oxidants in Europe. Calculations for July 1985. *Atmospheric Environment* **26A**, 1609–1634.

Simpson, D. 1993. Photochemical model calculations over Europe for two extended summer periods: 1985 and 1989. Model results and comparisons with observations. *Atmospheric Environment* **27A**, 921–943.

Simpson, D., Olendrzynski, K., Semb, A., Storen, E. and Unger, S. 1997. *Photochemical oxidant modelling in Europe: Multi-annual modelling and source-receptor relationships*. EMEP MSC-W Report 3/97.

Simpson, D., Fagerli, H., Jonson, J.E., Tsyro, S., Wind, P. and Tuovinen, J–P. 2003. *Transboundary acidification, eutrophication and ground level ozone in Europe*. Part 1. Unified EMEP model description. EMEP Report 1/2003.

Sinton, J.E., Smith, K.R., Hansheng, H. and Junzhuo, L. 1998. *Indoor air pollution database for China.* WHO Healthy Cities Air Management Information System AMIS 2.0.

Skeffington, R. 1999. The use of critical loads in environmental policy making: a critical appraisal. *Environmental Science and Technology* 33, 245A-252A

Smith, F.B. and Jeffrey, G.H. 1975. Airborne transport of sulphur dioxide from the UK. *Atmospheric Environment* 9, 643-659.

Smith, R.I., Fowler, D., Sutton, M.A., Flechard, C. and Coyle, M. 2000. A model for regional estimates of sulphur dioxide, nitrogen dioxide and ammonia dry deposition in the UK. *Atmospheric Environment* 34, 3757–3777.

Solomon, S. 1999. Stratospheric ozone depletion: a review of concepts and history. *Reviews of Geophysics* 37, 275–316.

SORG (Stratospheric Ozone Review Group) 1999. *Stratospheric ozone 1999.* DETR.

South Coast Air Management District 1997. *1997 Annual Report.* (http://www.aqmd.gov/pubinfo/97annual.html).

Staehelin, J., Renaud, A., Bader, J., McPeters, R.D., Vialte, P., Hoegger, B., Bugnion, V., Giraud, M. and Schill, H. 1998. Total ozone series at Arosa (Switzerland): homogenization and data comparison. *Journal of Geophysical Research* 108, 5827–5841.

Stedman, J. 2004. The predicted number of air pollution related deaths in the UK during the August 2003 heatwave. *Atmospheric Environment* 38, 1087–1090.

Stedman, J.R., Vincent, K.J., Campbell, G.W., Goodwin, J.W.L. and Downing, C.E.H. 1997. New high resolution maps of estimated background ambient NO_x and NO_2 concentrations in the UK. *Atmospheric Environment* 31, 3591–3602.

Stern, D.I. and Kaufmann, R.K. 1998. *Annual estimates of global anthropogenic methane emissions 1860–1994. Trends online: a compendium of data on global change.* Carbon Dioxide Information Analysis Center, Oak Ridge National Laboratory, Tennessee.

Stockwell, W.R., Middleton, P and Chang, J.S. 1990. The second generation regional acid deposition model chemical mechanism for regional air quality modelling. *Journal of Geophysical Research* 95 (D10), 16,343–16,367.

Stoddard, J.L., Jeffries, D.S., Lukewille, A., Clair, T.A., Dillon, P.J., Driscoll, C.T., Forsius, M., Johannessen, M., Kahl, J.S., Kellogg, J.H., Kemp, A., Mannio, J., Monteith, D.T., Murdoch, P.S., Patrick, S., Rebsdorf, A. Skelkvale, B.L., Stainton, M.P., Traaen, T., van Dam, H., Webster, K.E., Wieting, J. and Wilander, A. 1999. Regional trends in aquatic recovery from acidification in North America and Europe. *Nature* 401, 575–578.

Stolarski, R., Bloomfield, P., McPeters, R.D. and Harman, J. 1991. Total ozone trends deduced from Nimbus-7 TOMS. *Geophysical Research Letters* 18, 1015–1018.

Stouffer, R.J., Manabe, S. and Vinnikov, K. 1994. Model assessment and the role of natural variability in recent global warming. *Nature* 367, 634–636.

Street-Perrott, F.A. 1991. General circulation (GCM) modelling of palaeoclimates: a critique. *The Holocene* 1, 74–80.

Streets, D.G., Tsai, N.Y., Akimoto, H. and Oka, K. 2001. Trends in emissions of acidifying species in Asia 1985–1997. *Water, Air and Soil Pollution* 130, 187–192.

Streets, D.G., Bond, T.C., Carmichael, G.R., Fernandes, S.D., Fu, Q., He, D., Klimont, Z., Nelson, S.M., Tsai, N.Y., Wang, M.Q., Woo, J-H. and Barber, K.F. 2003. An inventory of gaseous and primary aerosol emissions in Asia in the year 2000. *Journal of Geophysical Research Atmospheres* 108 (D21), article number 8809.

Streit, G.E. and Guzman, F. 1996. Mexico City air quality: progress of an international collaborative project to define air quality management options. *Atmospheric Environment* 30, 723–733.

Sturges, W.T., Wallington, T.J., Hurley, M.D., Shine, K.P., Sihra, K., Engel, A., Oram, D.E., Penkett, S.A., Mulvaney, R. and Brenninkmeijer, C.A. 2000. A potent greenhouse gas identified in the atmosphere: SF_5CF_3. *Science* 289, 611–613.

Suess, H.E. 1955. Radiocarbon concentration in modern wood. *Science* 122, 415–417.

Sundet, J.K. 1997. *Model studies with a 3-D CTM using ECMWF data.* PhD thesis, Department of Geophysics, University of Oslo, Norway.

Sundqvist, G., Letell, M. and Lidskog, R. 2002. Science and policy in air pollution abatement strategies. *Environmental Science and Policy* 5, 147–156.

Sutton, M., Place, C.J., Eager, M., Fowler, D. and Smith, R.I. 1995. Assessment of the magnitude of ammonia emissions in the United Kingdom. *Atmospheric Environment* 29, 1393–1411.

Tao, M-S., and Del Genio, A. 1999. Effects of parameterisation on the simulation of climate changes in the GISS GCM. *Journal of Climate* 12, 761–779.

Tarrason, L., Semb, A., Hjellbrekke, A-G., Tsyro, S., Schaug, J., Bartnicki, J. and Solberg, S. 1998. *Geographical distribution of sulphur and nitrogen compounds in Europe derived both from modelled and observed concentrations.* EMEP/MSW-W Note 4/98.

Tarrason, L., Fagerli, H., Hjellbrekke, A-G., Ukkelberg, A. and Posch, M. 2001. *Transposition from Lagrangian to Eulerian models in EMEP. In: Transboundary acidification, eutrophication and ground level ozone in Europe.* EMEP summary report 2001. EMEP Report 1/2001, chapter 3.

The Ministry of the Environment of the Czech Republic 2001. *The Czech Republic's Third National Communication on the UN Framework Convention on Climate Change.* Czech Hydrometeorological Institute.

The Smithsonian 1989. *The Smithsonian Guide to Historic America. The Pacific States.*

The World Commission on Environment and Development 1987. *Our Common Future.* OUP.

Thompson, L.G., Moseley-Thompson, E., Davis, M.E., Lin, P-N., Henderson, K.A., Cole-Dai, J., Bolzan, J. and

Liu, K-b. 1995. Late glacial stage and Holocene tropical ice core records from Huascaran, Peru. *Science* **269**, 46–50.

Thompson, L.G., Yao, T., Davis, M.E., Henderson, K.A., Mosley-Thompson, E., Lin, P-N., Beer, J., Synak, H-A., Cole-Dai, J. and Bolzan, J.F. 1997. Tropical climate instability: the last glacial cycle from a Qinghai-Tibetan ice core. *Science* **276**, 1821–1825.

Thompson, A.M., Witte, J.C., Hudson, R., Guo Hua, Herman, J.R. and Fujiwara, M. 2001. Tropical tropospheric ozone and biomass burning. *Science* **291**, 2128–2132.

Tsyro, S. 1998. *Description of the Lagrangian Acid Deposition Model.* In: *Transboundary Acidfying Air Pollution in Europe.* EMEP/MSC-W Report 1/98 Part 2: Numerical Addendum.

Turco, R.P. 1997. *Earth under siege. From air pollution to global change.* Oxford University Press.

UNECE 1998. *Protocol to the 1979 Convention on Long Range Transboundary Air Pollution of Heavy Metals.* (Available at http://www.unece.org/env/lrtap/hm_h1.htm)

UNECE 1999. *Protocol to the 1979 Convention on Long Range Transboundary Air Pollution to abate acidification, eutrophication and ground level ozone.* (Available at http://www.unece.org/env/lrtap/multi_h1.htm)

UNEP 1994. *Environmental effects of ozone depletion: 1994 assessment.* Also published as a special issue of *Ambio*, **24** (3) (1995)

UNEP/WHO 1992. *Urban air pollution in megacities of the world.* Blackwell, Oxford.

UNEP/WHO 1996. *Air quality management and assessment capabilities in 20 major cities.* UNEP, Nairobi/WHO, Geneva.

Van Aardenne, J.A., Dentener, F.J., Olivier, J.G., Klein Goldwijk, C.G. and Lelieveld, J. 2001. A 1 x 1 degree resolution dataset of historical anthropogenic trace gas emissions for the period 1890 – 1990. *Global Biogeochemical Cycles* **15**, 909–928.

Vandal, G.M., Fitzgerald, W.F., Boutron, C.F. and Candelone, J-P. 1995. *Mercury in ancient ice and recent snow from the Antarctic.* NATO ASI series vol. 130, 401–415.

Vestreng, V. and Klein, H. 2002. *Emissions data reported to UNECE/EMEP: quality assurance and trend analysis and presentation of WebDab.* EMEP/MSC-W Note 1/2002.

Vestreng, V. 2003. *Review and revision. Emission data reported to CLRTAP.* MSC-W Status Report. EMEP/MSC-W Note 1/2003.

Vettoretti, G., Peltier, W.R. and McFarlane, N.A. 1998. Simulations of mid-Holocene climate using an atmospheric general circulation model. *Journal of Climate* **11**, 2607–2627.

Washington, W.M., Weatherly, J.W., Meehl, G.A., Semtner, A.J., Bettge, T.W., Craig, A.P., Strand, W.G., Arblaster, J.M., Wayland, V.B., James, R. and Zhang, Y. 2000. Parallel climate model (PCM) control and transient simulations. *Climate Dynamics* **16**, 755–774.

Watterson, I.G., Dix, M.R. and Colman, R.A. 1998. A comparison of present and doubled CO_2 climates and feedbacks simulated by three general circulation models. *Journal of Geophysical Research* **104**, 1943–1956.

Wellburn, A. 1994. *Air pollution and climate change. The biological impact.* Longman Scientific and Technical. 2nd edition.

Whelpdale, D.M. and Kaiser, M.S. 1996. *Global acid deposition assessment.* WMO, Global Atmosphere Watch No. 106.

WHO 1987. *Air quality guidelines for Europe.* World Health Organisation Regional Publications, European Series no. 23. Regional Office for Europe, Copenhagen, Denmark.

WHO 2000. *Guidelines for Air Quality.* WHO, Geneva. (http://www.who.int/peh/)

WHO 2002. *The world health report 2002. Reducing risk, promoting healthy life.* (http://www.who.int/whr/2002)

WHO 2003. INTERSUN. *The global UV project. A guide and companion to reduce the burden of disease resulting from exposure to UV radiation while enjoying the sun safely.* (http://www.who.int/uv/publications/en/intersunguide.pdf).

Williams, K.D., Senior, C.A. and Mitchell, J.F.B. 2001. Transient climate change in the Hadley Centre models: The roles of physical processes. *Journal of Climate* **14**, 2659–2674.

Wind, P., Tarrason, L., Berge, E., Havard Slørdal, L., Solberg, S. and Walker, S-E. 2002. *Development of a modeling system able to link hemispheric-regional and local air pollution.* EMEP Note 5/2002.

WMO 1989. *The atmospheric input of trace species to the world ocean.* GESAMP Report and Studies No. 38. WMO, Geneva.

WMO 1999. *Scientific assessment of ozone depletion: 1998.* Global Ozone Research and Monitoring Project, Report No. 44.

WMO 2003. *Scientific assessment of ozone depletion: 2002.* Global Ozone Research and Monitoring Project, Report No. 47. World Meteorological Organisation.

Wolff, E.W. 1995. *Nitrate in polar ice.* NATO ASI Series vol. 30, 195–224.

Wolff, G.T., Lioy, P.J., Wright, G.D., Meyers, R.E. and Cederwall, R.T. 1977. *An investigation of long range transport of ozone across the Midwestern and Eastern United States.* In: *Proceedings of an International Conference on Photochemical Oxidant Pollution and its controls.* EPA-600/3-77-001a. Environmental Sciences Research Laboratory, North Carolina, USA.

Wright Jr., H.E., Kutzbach, J.E., Webb III, T., Ruddiman, W.F., Street-Perrott, F.A and Bartlein, P.J. eds. 1993. *Global climates since the last glacial maximum.* University of Minnesota Press.

Wuster, H. 1992. *The convention on long-range transboundary air pollution: Its achievements and its potential.* In: *Acidification Research: Evaluation and Policy Applications.* **Edited by T. Schneider**, pp. 221–239. Elsevier, Amsterdam, the Netherlands.

WWICS 1996. *Summary of working group discussions on energy issues in China*. China Environment Series, Woodrow Wilson International Center for Scholars.

WWICS 1997. *The Chinese political economy and central-local government dynamics urban, township and village air pollution*. China Environment Series, Woodrow Wilson International Center for Scholars.

Xenopoulos, M.A., Prairie, Y.T. and Bird, D.F. 2000. Influence of ultraviolet-B radiation, stratospheric ozone variability, and thermal stratification on the phytoplankton biomass dynamics in a mesohumic lake. *Canadian Journal of Fisheries and Aquatic Sciences* **57**, 600–609.

Yamartino, R.J., Scire, J.S., Carmichael, G.R. and Chang, Y.S. 1992. The CALGRID mesoscale photochemical grid model. – I. Model formulation. *Atmospheric Environment* **26A**, 1493–1512.

Yao, M.S. and Del Genio, A.D. 1999.

Yukimoto, S., Noda, A., Kitoh, A., Sugi, M., Kitamura, Y., Hosaka, M., Shibata, K., Maeda, S. and Uchiyama, T. 2001. A new meteorological research institute coupled GCM (MRI-CGCM2) – model climate and its variability. *Papers in Meteorological Geophysics* **51**, 47–88.

Zepp, R., Callaghan, T. and Erickson, D. 1995. Effects of increased solar ultraviolet radiation on biogeochemical cycles. *Ambio* **24**, 181–187.

Zielinski, G.A., Mayewski, P.A., Meeker, L.D., Whitlow, S., Twickler, M.S., Morrison, M., Meese, D.A., Gow, A.J. and Alley, A.B. 1994. Record of volcanism since 7000-BC from the GISP2 Greenland ice core and implications for the volcano-climate system. *Science* **264**, 948–952.

Zielinski, G.A. 2000. Use of paleo-records in determining variability within the volcanism-climate system. *Quaternary Science Reviews* **19**, 417–438.

Index

Note: For reasons of space, items that are generally referred to by abbreviations or acronyms are entered under these shortened forms only (e.g. CFCs). However, chemicals are entered under names rather than chemical symbols. Page numbers in *italics* refer to figures and tables.